The Synthesis of Three Dimensional Haptic Textures: Geometry, Control, and Psychophysics

Gianni Campion
Montreal
Canada

ISSN 2192-2977 e-ISSN 2192-2985
ISBN 978-1-4471-2654-6 ISBN 978-0-85729-576-7 (eBook)
DOI 10.1007/978-0-85729-576-7
Springer London Dordrecht Heidelberg New York

British Library Cataloguing in Publication Data
A catalogue record for this book is available from the British Library

Cover design: deblik

Printed on acid-free paper

Springer is part of Springer Science+Business Media (www.springer.com)

To Elena

Series Editors' Foreword

Haptics is a multi-disciplinary field with researchers from Psychology, Physiology, Neurology, Engineering, and Computer Science (amongst others) that contribute to a better understanding of the sense of touch, and research on how to improve and reproduce haptic interaction artificially in order to simulate real scenarios.

The "*Springer Series on Touch and Haptic Systems*" is a new *Springer* book series published in collaboration with the EuroHaptics Society. It is focused on publishing new advances and developments in all aspects of haptics. The goal is to obtain a fast dissemination of the latest results in order to stimulate the interaction among members of the haptics community and to promote a better understanding of touch perception and find the most suitable technologies to reproduce and simulate haptic environments.

The first issue of this series has been prepared by Gianni Campion, and is based on his PhD thesis. The content is focused tactile texture perception, a highly relevant topic in the field of haptics, and covers the simulation of textures and their evaluation with psychophysical methods.

The selection of this thesis for publication reflects the interest in the topic of texture perception and the high quality of the work. Being a thesis, it covers the topic in a very focused manner and analyzes it in considerable depth. As series editors we will continue to encourage this kind of publication as well as supporting publication of books focused on more general topics.

Finally, the series editors would like to thank the EuroHaptics Society for promoting haptics and for supporting this exciting new book series by Springer on Touch and Haptic Systems. Moreover, we would also like to thank all the members of the Series Editorial Advisory Board for their contributions in reviewing and so ensuring high quality of the publications.

Manuel Ferre
Marc O. Ernst
Alan Wing

Series Editor's Foreword

Foreword

"The Synthesis of Three-Dimensional Haptic Textures: Geometry, Control and Psychophysics" by Gianni Campion under the advisement of Dr. V. Hayward presents a series of innovative tools that can be used to remove the artifacts from haptic rendering of textures. The main contributions include a complete platform, device, and synthesis algorithm, as well as evaluation of the techniques.

Overall, this book presents an all-front attack and very in-depth investigation of all components involved in haptic rendering of textures: hardware, software and psychophysics. The proposed techniques are effective and clever. I have worked in these areas for over a decade. There is a huge collection of literature in all these areas. I'm impressed that the work has done an excellent effort in surveying prior research, analyzing previous work, proposing new points of view, and synthesizing techniques to improve the overall rendering performance of haptic textures. The technical writing of the book is clear, coherent, carefully thought-out and well-organized. The diagrams and captured images clearly illustrate the basic concepts and further enhance the overall presentation. I believe the findings and results would be of significant interest to the haptics and robotics community.

Chapel Hill
December 2010

Ming Lin

Foreword

Working with Gianni Campion has been a most gratifying experience. Gianni started out as a self proclaimed computer scientist who would not even touch a screwdriver with a six-foot pole, but ended up having fun in the workshop making (simple) parts with the lathe more often than he would care to confess. The results of his voracious intellectual curiosity are evident throughout his work which is a must-read for anyone interested in haptic virtual environments where the surfaces are, as they should be, not smooth.

Gianni, again, congratulations for a job well done.

Paris
December 29, 2010

Vincent Hayward

Acknowledgements

I would like to thank Prof. Vincent Hayward for his kind supervision, his willingness to share his (numerous) ideas and insights, and for his generous style of teaching.

My colleagues in the Haptics Laboratory were always open to discuss the most various topics, the majority of which were not even loosely related to this thesis. I would like thank them in random order: Andrew Gosline with his magnets, Qi Wang, Hsin-Yun Yao and the PCBS, Mohsen Mahvash, Vincent Levesque the coder, Jerome Pasquero, Hanifa Dostmohamed, Omar Ayoub, Mounia Ziat, and Diana Garroway. I would not dare to forget the support of the people at the Center for Intelligent Machines, specially Cynthia Davidson, who has been a seamless interface with the bureaucratic side of McGill, and Jan Binder, who answered the too many requests I had for the System Administrator.

This research was supported in part by the Institute for Robotics and Intelligent Systems, in part by NSERC the Natural Sciences and Engineering Research Council of Canada, and by Immersion Corp. I would also like to acknowledge the reception of a PRECARN Inc. Scholarship, a McConnell McGill Major Scholarship, and a CGS-D2 Scholarship from NSERC.

Finally I thank my family for their support to this endeavor and Elena, who helped me through this effort.

Acknowledgements

Contents

Chapter 1
Introduction

Abstract This chapter introduces the main topics discussed in the book and defines the scope of the research presented. Specifically, this book discusses the rendering of haptics textures with force-feedback haptic devices and takles the topic both from the engineering and the psychophysics angle.

1.1 Introduction

Human touch is a versatile sense. It is used to explore the environment, as a control mechanism for movement and manipulation, and even as a non-verbal communication channel; as an example, visually impaired people may rely on touch for reading. The discipline which studies the sense of touch is called *haptics* (from the Greek *haptô, hapsasthai*); the term *haptic* is an adjective meaning "Of or relating to the sense of touch" [1].

Despite the flexibility of the sense of touch, the development and availability of haptic interfaces greatly lags behind that of visual interfaces (e.g., monitors and TV) and audio technology (loudspeakers and headphones). In fact, the basic nature of tactile sensation is still under investigation. While visual stimuli are known to be electromagnetic radiations of certain wavelengths and audio stimuli are pressure waves reaching the eardrum, the basic nature of the haptic stimuli is yet to be fully understood. This lack of fundamental knowledge about the sense of touch is compounded by the lack of haptic devices capable of delivering controlled stimuli as rich as the contact interactions between the skin and the surface of an object. The divide between the natural stimuli and artificial equivalent is particularly pronounced when generating virtual textures, because of their significant high frequency components.

There are two main modalities of haptic interaction with objects: direct touch exploration requires the contact of the skin (usually a finger) with the object, the second is indirect touch, where the skin contacts a proxy and the proxy scans the object. Delivering controllable textured stimuli for bare finger exploration is extremely complex and, at the time of writing, very few attempts have been made with mixed results. The most daunting problem is the spatial resolution of the textures which can be resolved by touch. Humans can perceive textural elements less than 200 μm apart, and delivering a controlled deformation to the skin at that scale is still not feasible. More encouraging results are obtained for indirect touch, where

G. Campion, *The Synthesis of Three Dimensional Haptic Textures: Geometry, Control, and Psychophysics*, Springer Series on Touch and Haptic Systems,
DOI 10.1007/978-0-85729-576-7_1, © Springer-Verlag London Limited 2011

the user interacts with a surface through a proxy; but also in this case, the human somatosensory system can detect and discriminate stimuli to a level which cannot be attained by currently available proxy-based haptic devices.

Moreover, a single haptic device cannot render all the possible force signals, the same way a visual display cannot produce every possible visual stimulus. For example, the spatial resolution of a computer screen limits the size of the smallest feature displayable and the frequency of the spatial variations of light. Similar limitations occur in haptic devices and a framework for assessing the effects of those limitations is needed.

To compound this problem, the algorithms presented in the literature are discussed only in relation to their psychophysical properties, but their energy profile is never characterized, nor a formal passivity-based analysis is performed. As a result, it is extremely difficult to interpret the psychophysical results reported and it is impossible to extend those findings to haptic devices different from the one used in the specific example.

1.2 Scope

This book focuses on the problem of generating force-feedback textures precisely and free of artifacts. Force-feedback is understood to refer to the most common approach adopted to create touch sensations in virtual reality settings. Users "touch" a virtual environment through an electromechanical device acting like an intermediary [2]. The feeling of touching a virtual object is generated by varying the force acting on the proxy in response to the user motion.

This book deals with both haptic devices and the rendering algorithms. Regarding the former, it presents a set of conditions highlighting the sources of artifacts due to the haptic devices. Texture algorithms, on the other hand, are explored with a novel analytic tool derived from passivity theory that removes the imperfections of the rendering due to energy imbalance. This framework is used to validate a rendering platform (device and algorithm) which can be used to explore the perception of haptic textures. In particular, a psychophysical experiment aimed at investigating the equivalence between texture algorithms with regard to the roughness perception elicited is presented.

1.3 Overview

The book is organized in 10 chapters: this introduction, a literature review, five manuscripts, two chapters, and a summary.

Chapter 2 covers the previous work in the domain of haptic textures. It contains an overview of the most relevant haptic devices, a comprehensive list of the texture algorithms developed for force-feedback haptic textures, a survey of the major results in control applied to haptics (particularly the passivity analysis and the resulting conditions of virtual environments), a review of the psychophysics of texture perception (both for real and virtual textures) as well as a brief summary of the physiology of touch.

Chapter 3 describes the properties of the re-engineered Pantograph haptic device, which was used in the rest of the book to implement, test, and validate the properties of virtual texture algorithms. The most notable part of this chapter is the oversample and filter approach. By pushing the sampling rate to 10 kHz and by filtering the torque commands generated by the texture algorithm, the force signal at the finger becomes extremely clean, and free from sampling artifacts.

The purpose of this chapter was to introduce the Pantograph as an "open architecture" haptic device, which could then be used as a standard reference; in the process, however, it became clear that the Pantograph was extremely well suited for rendering haptic textures, for its high position resolution and the large acceleration bandwidth.

Chapter 4 answers to the lack of framework identified in the literature review. Six conditions are proposed in this chapter, covering the problems of resolution, spatial and temporal aliasing, force quantization and passivity margins. Five of those conditions describe the rendering capabilities of a haptic device, while the last one is more intimately related to the rendering algorithm.

The second part is spent to confirm the qualities of the Pantograph with an experiment; the acceleration measured at the tooltip of the device has a frequency spectrum close to the desired force signal. In the end, the re-engineered Pantograph is an ideal testbed for research on haptic textures, because it is the first haptic device whose frequency response is adequate for synthesizing haptic textures up to 400 Hz.

Chapter 5 investigates the energy properties of different haptic texture algorithms; to summarize the effects of a texture algorithm on the passivity of the haptic interaction, a novel measure is introduced, called the characteristic number. This new tool offers numerous insights on the parameters of the textures algorithms; for example, it can explain the instabilities found by previous authors when using bump mapping techniques.

A second application of the characteristic number is to ensure the passive rendering of haptic textures, to avoid the typical "buzzing vibrations" generated by virtual environments. Once the passive rendering is formally guaranteed, the artifacts intrinsic to the haptic algorithm can be investigated. For example, non-conservative force fields are shown to be affected by the so-called "aliveness" artifact, although the haptic interaction might be locally passive. Finally, a novel formulation for a friction based texture algorithm is formally proposed and analyzed.

In Chap. 6 the theory of passivity for non linear and multidimensional virtual environments is extended to address the problem of spatial quantization. This analysis contributes to the understanding of the interactions between algorithms and haptic devices, and confirms the validity of the characteristic number also in presence of non negligible spatial quantization.

Here, the notion of *passive realization* is introduced, to extend the passivity analysis to non-conservative force fields, which are a common occurrence in haptic textures. In the literature, there is no mention of the distinction between conservative and non-conservative force fields.

Chapter 7 generalizes the characteristic number to generic 3D curved surfaces. It contains two notable results for algorithms based on the normal penetration in a curved surface. First, these algorithms can suffer from severe and localized lack of passivity for convex surfaces. Second, the apparent pitch of the texture is distorted as a function of curvature of the surface and penetration.

Chapter 8 and Chap. 9 explore the psychophysics of the novel algorithm for haptic textures based on friction.

First an investigation of the perceptual space generated by varying the friction coefficient, pitch, and amplitude of sinusoidal gratings is carried out. It was found that roughness scales monotonically with the lateral force variations when textures have the same pitch.

Based on this result, a fast calibration method for haptic textures is implemented, and the friction based algorithm is shown to generate a roughness sensation equivalent to a geometric based algorithm.

This experiment is the first successful attempt to calibrate two different texture algorithms based on the percept of roughness. When the roughness of the two algorithms is matched, the characteristic number can be used to fairly assess the passivity margins of the resulting haptic interaction.

Chapter 10 concludes the book with a summary and a discussion of the major findings.

The innovative aspects of this work regard the engineering properties of haptic devices, virtual environments, and specifically haptic textures; nevertheless, a psychophysical investigation is required to contextualize the passivity properties of virtual textures.

The two experiments reported here fulfill this duty by finding the perceptual equivalence of two texture algorithms which can be then compared with respect to their passivity margins. This last step hints a different use of the characteristic number, which is now a fair tool for comparing different algorithms based on their passivity margins.

1.4 Summary of Contributions

The book contains the following contributions:

- The redesign of the digital controller of the Pantograph haptic device, resulting in a force-feedback device capable of rendering textures up to 400 Hz. The Pantograph is also thoroughly analyzed to confirm that the minimal specifications for texture synthesis are met.
- A framework of six conditions identifying the most common sources of rendering artifacts. These conditions are mostly related to the hardware properties of the haptic device.
- The analysis of the effects of spatial quantization on the passivity margin of multidimensional, non-linear virtual environments; and the new concept of "passively realizable" to extend passivity properties to non-conservative virtual environments, which are by definition non passive.

- The definition and validation of a measure for the passivity margins of virtual haptic textures rendering. This measure, called characteristic number, is applied to both 1D flat textures as well as 3D curved, textured surfaces.
- The analysis of currently available texture algorithms, with respect to the characteristic number and the energy profile (conservative, dissipative, generative). This theoretical investigation confirms, for the first time, the existence of two distinct classes of artifacts, arising from the lack of passivity and from the energy profile of the texture algorithm.
- A novel formulation of a friction based texture algorithm and its characterization both over 1D flat virtual walls and over 3D curved surfaces. A psychophysical method that can establish the perceptual correspondence between the parameters of this novel algorithm with those of other algorithms.
- A psychophysical study of virtual texture roughness. Based on a fast calibration method, this study shows that, in general, two different texture algorithms can be matched for roughness. Once the point of subjective equivalence of two algorithms is found, it is possible to compare their passivity margin fairly.

References

1. Colman, A.M.: A Dictionary of Psychology. Oxford University Press, Oxford (1996). Oxford Reference Online, 21 October 2007. http://www.oxfordreference.com/
2. Hayward, V., Maclean, K.E.: Do it yourself haptics: Part I. IEEE Robot. Autom. Mag. **14**(4), 88–104 (2007)

Chapter 2
Literature Review

Abstract This chapter presents a literature review of the previous work related to haptic textures. After an overview of the most relevant devices, control strategies, and algorithms used in haptics, the author presents the major findings on the perception of haptic textures and roughness. This review covers both the psychophysics experiments as well as the basic results of the physiology of tactile perception of textures and surfaces. The chapter is concluded with the discussion of the current understanding of the perception of virtual haptic textures generated with force feedback devices, thus setting the stage for the discussion of the research presented in the following chapters.

2.1 Introduction

The discussion of novel techniques and solutions regarding the rendering of haptic textures requires some background material on haptic devices, Sect. 2.2 at page 7; on algorithms used to generate virtual haptic sensation, Sect. 2.5 at page 29; on control theory applied to haptics, Sect. 2.3 at page 14; and on the psychophysics behind the haptic perception of textures, Sect. 2.4 at page 19. These four topics belong to the areas of engineering, psychophysics, and physiology, which are the core disciplines contributing to haptic technologies and that are intimately related. Without a precise characterization of the haptic device used, it is impossible to properly interpret any psychophysical experiment conducted on virtual haptic sensations, and it might be difficult to extend findings gained from one device to other devices. At the same time, facts related to the human somatosensory system (obtained with classical psychophysical studies on real tactile stimuli) offer valuable guidelines for improving the design of applications.

The centerpiece of any computer controlled haptic interaction is clearly the haptic device, which is the hardware used to stimulate the somatosensory system of the user.

2.2 Interfaces

In general, a haptic device, which is also called a haptic display, resembles a robotic system that applies computer generated tactile stimulation to the skin of a human

G. Campion, *The Synthesis of Three Dimensional Haptic Textures: Geometry, Control, and Psychophysics*, Springer Series on Touch and Haptic Systems, DOI 10.1007/978-0-85729-576-7_2, © Springer-Verlag London Limited 2011

user. These devices can be roughly categorized in four classes: vibrotactile, surface, tactile, and force-feedback interfaces.

Vibrotactile displays are designed to apply high frequency and low amplitude vibrations to the skin; stimulations can be applied to the hand as well as to other body parts, depending on the application. Vibrotactile stimulators are very common and can be found in controllers for video games and in cell phones; the key to their availability is the low cost of the actuators, usually a motor with an eccentric mass, but their capabilities are extremely limited.

Conversely, tactile displays produce a distributed deformation pattern to the fingerpad, either through indentation or stretching of the skin. Tactile interfaces can be used, for example, to provide blind users with access to digital media. Refreshable braille cells can be assembled to display a single line of characters from a text file. Recent advances in tactile simulators suggest the possibility of refreshable tactile graphics based on skin stretch [93].

Surface displays are based on the observation that the area of contact of the finger with an object changes during the exploration process. The actuators in the device move a flat plate against the finger of the user to change the size and location of the area of contact, producing the sensation of a shape. By changing the control of the device it is possible to produce convex, concave, and flat surfaces.

Finally, force-feedback devices can deliver a single, low frequency force signal to the user's hand, usually through a pen-like interface, a knob, or a thimble. This class of devices has found a niche application in virtual reality, where they provide programmable haptic stimulation to the user. Among the most significant applications that benefit from force-feedback are: surgical simulation, virtual sculpture, CAD modeling, remote sensing, and video games. Although commercial devices offer solutions for each of these cases, it is difficult to use them in a research environment due to their inherent limitations. Most importantly, the study of virtual textures requires a degree of fidelity that is yet to be achieved in general purpose devices.

Only force-feedback devices relevant to the scope of this book are discussed. Before analyzing the different devices available and their characteristics, it is important to review some basic rendering algorithms for virtual environments.

2.2.1 Virtual Environments

Given the Cartesian position of a manipulandum $\mathbf{x} = [x_1\ x_2\ x_3]^\mathsf{T}$, a rendering algorithm computes a force field $\mathbf{F}(\mathbf{x}, \dot{\mathbf{x}})$ over the workspace. In general, virtual environments are defined as combinations of some fundamental elements, such as elasticity, viscosity, and friction.

For example, elastic unilateral constraints can simulate the boundary of objects; in 3D space it is possible to implement a so-called virtual wall in the half volume

$x_3 < 0$ with the field:

$$\mathbf{F}_\mathrm{K}(\mathbf{x}, \dot{\mathbf{x}}) = \begin{cases} [0, 0, -K_\mathrm{V} x_3]^\mathrm{T} & \text{if } x_3 < 0, \\ [0, 0, 0]^\mathrm{T} & \text{otherwise}, \end{cases} \tag{2.1}$$

where K_V is the stiffness of the virtual wall. Viscous friction (also called damping) improves the quality of the rendering of virtual walls and can be synthesized according to:

$$\mathbf{F}_\mathrm{B}(\mathbf{x}, \dot{\mathbf{x}}) = \begin{cases} [0, 0, -B_\mathrm{V} \dot{x_3}]^\mathrm{T} & \text{if } x_3 < 0, \\ [0, 0]^\mathrm{T} & \text{otherwise}, \end{cases} \tag{2.2}$$

where B_V is the damping coefficient; the virtual wall response in the same half volume of Eq. (2.1) is computed as

$$\mathbf{F}_\mathrm{W}(\mathbf{x}, \dot{\mathbf{x}}) = \mathbf{F}_\mathrm{K}(\mathbf{x}, \dot{\mathbf{x}}) + \mathbf{F}_\mathrm{B}(\mathbf{x}, \dot{\mathbf{x}}). \tag{2.3}$$

The relevance of damping will be clear when discussing the passivity framework for haptic devices.

Virtual stiffness and damping are simple effects that can be summarized in a single equation; on the contrary, dry friction is a phenomenon that cannot be captured by a simple formula. Different rendering algorithms have been proposed for virtual friction, and a specific one will be discussed in relation to the new rendering approach introduced in this book. Informally, dry friction is a force arising at the contact of two surfaces and opposes their relative motion independently from the speed of the motion.

These elements of virtual environments are useful to analyze the performance of force feedback devices: it is possible to compare different devices based on the maximum stiffness and damping they can render.

2.2.2 Force Feedback Devices

Force feedback devices are robotics mechanisms (composed of a mechanical structure, actuators, sensors, and a control unit) which can exchange energy with the hand of a user through a handle (for example a stylus, a knob, or a thimble).

2.2.2.1 Generalities

There are two classes of force feedback devices: admittance displays and impedance displays.

Admittance devices measure the force $\mathbf{F}$ that the user is exerting on the handle and respond by moving the end effector according to a rendering law $\left[\begin{smallmatrix}\mathbf{x}(\mathbf{F})\\ \dot{\mathbf{x}}(\mathbf{F})\end{smallmatrix}\right]$. Usually admittance devices must be strong because they need to impose a precise trajectory to the end effector in response to user's force. The frequency response and position accuracy of such devices are typically not suitable for haptic textures because they rely on gears and transmissions to provide large forces to the user, but they can mask small displacements due to friction.

Impedance devices are more attractive for haptic texture rendering because they designed to have little inherent friction, damping, and mass. These characteristics contribute to the transparency of the interface and the user is less aware of the device and can focus more on the virtual environment. The rendering algorithms in Sect. 2.2.1 at page 8 are expressed for a impedance style interface: a force field $\mathbf{F}$ is displayed as a function of the position of the device $\mathbf{x}$ and velocity $\dot{\mathbf{x}}$. In general, these devices can render forces much weaker than admittance devices, but the frequency characteristics of the rendering is much more suitable for haptic textures.

Thus, only impedance style devices will be considered for the scope of this work.

2.2.2.2 Force/Torque Output Capabilities of the Device

During the rendering of a virtual environment a force feedback device acquires the position and/or orientation of the handle held by the user, which results in an N dimensional vector; it then generates forces and/or torques which span an M dimensional space in response to the user actions. The rendering algorithm computes the force $\mathbf{F}_i$ and the torques $\boldsymbol{\tau}_i$ at time step i as a function of possibly all the previous position readings $\mathbf{x}_j : 0 < j \leq i$; generally, the only the last measured position $\mathbf{x}_i$ and the last estimate of the handle velocity $\hat{\dot{\mathbf{x}}}_i$ are used to compute the forces.

It is possible to simplify the analysis by categorizing the devices according to dimension of the force/torque space M; in this book only force feedback devices relevant to haptic textures are presented. A notable omission are exoskeleton type devices, which can deliver multiple torques to different joints of the arm and wrist, with the goal of rendering very large objects. These devices are not used to generate force feedback textures because of their performance, for example a backlash of 1 cm is unacceptable for haptic texture rendering, but it is more than acceptable for exoskeleton type devices [41].

2.2.2.3 1D Torque or Force

The haptic knob is the simplest force feedback device available, see for example [69, 96]. It consists of a single rotary actuator, an angular sensor, and a small wheel through which the torques are transferred to the user. The actuator can be either a brake (which gives a resistive torque) or a motor (which provides active torques). In both cases, haptic textures are rendered by varying the torque output as a function of the rotational position and angular velocity of the device; this rendering can result in the user feeling a textured knob. In a more commercial application, a haptic knob can simulate detents typical of the controls of hi-fi systems, thus providing a programmable controller for a stereo. Devices in this class, however, cannot render virtual constraints and textures at the same time, because of the single degree of force output; nevertheless, haptic knobs represent a valuable tools for basic haptic research. The same architecture can be used for implementing the force feedback steering wheels used in arcade driving games, [94].

A variation on the haptic knob is the haptic paddle that converts the torque of a single actuator in a force, displayed to the user through a handle, [124]. This device suffers from the same limitations of the knob but has been extensively involved in experiments aimed at validating the control theory results pertinent to haptic devices, e.g. [1].

2.2.2.4 2D Forces

Because they are able to render superficial properties, two dimensional haptic devices offer a good trade off between simplicity and capabilities. Usually based on parallel mechanical structures they can be divided in two main categories: joysticks, [67, 68, 103, 125], and planar devices, such as the Pantograph [11, 122]. The basic element is usually a five bar linkage, either spherical for joysticks or planar for Pantograph style devices. Due to the properties of five bar linkages, it is possible to have direct drive connection between the motors and the linkage while maintaining the stators grounded; this greatly reduces the apparent mass of the device without resorting to cable transmission.

Joystick-like devices have a stick that can rotate in 2 dimensions, pivoting through a fixed point; the actuators provide torques in the same two directions of rotation, those torques are felt as forces by the user holding the stick. Their rendering capabilities are limited to 2D force fields which depend on the orientation and angular velocity of the stick; as a result, the virtual boundaries that can be rendered are limited; for example, flat boundaries must contain the pivoting point. Nevertheless, joysticks are capable of generating the sensation of exploring 2D textured surfaces. Several researchers used the Immersion Impulse Engine 2000 joystick for haptic texture rendering, as discussed in Sect. 2.5 at page 29.

The Pantograph haptic display, on the other hand, renders a planar field of 2D forces that are felt through a small plate contacting a single finger. The plate stimulates both the kinesthetic and the tactile channel because the forces deform the fingerpad and move the finger at the same time. This device is used implement both 2D varying force fields as well as 1D planar constraints on which textures can be applied. It is then possible to use such devices for investigating almost all the rendering algorithms available for haptic textures. Since it is a device designed to render forces at the finger scale, the 2 N maximum force is sufficient for conveying realistic sensation. Stronger forces are necessary when dealing with arm-size devices. With regard to texture research, West and Cutkosky realized a custom-made 2D device to compare the detection of haptic textures when exploring real and virtual surfaces in the same conditions and through the same interface [147].

2.2.2.5 2D Forces, 1D Torques

An intermediate step between 2D and 3D devices was proposed by Sirouspour et al., who realized a modified Pantograph design for rendering 2D planar forces and 1D torques perpendicular to the same plane, [133]. In their design, the robot acquires the position of the handle on a plane as well as its rotation around the axis perpendicular to the workspace, from which the force/torque commands are generated. This extension allows the simulation of planar interactions between rigid bodies and has both a large workspace and allows infinite rotations around the vertical axis, but has not yet been used for rendering virtual textures. This device is currently being manufactured by Quanser [121].

2.2.2.6 3D Forces

3D impedance-style force feedback haptic devices have been subject of extensive research and design, because they can be applied to a broad range of scenarios while maintaining a relatively simple design. With 3D force feedback devices it is possible to render 2D virtual boundaries with a superimposed texture.

The most successful 3D haptic device is the Sensable PHANTOM™; developed by Massie and Salisbury [99]. The device has three motors, which provide the three torques necessary to produce a full 3D force display; two motors act on a 5-bar linkage which provides forces on a plane, which is then rotated by the third motor to generate an arbitrary 3D force. Due to a careful distribution of the masses and to a capstan transmission system, the PHANTOM™ is almost balanced statically (hence the user does not support most of the weight of the device) and provides a nicely damped, strong force feedback feeling. Multiple versions have been produced with different sizes, strength, and materials: the high quality models, Premium 1.0 and Premium 1.5, are very interesting for haptic research, while low cost plastic versions are more suitable for less precise applications [131]. In general, the users interact with the PHANTOM™ by holding a stylus, at the end of which forces are applied.

Two of the motors of the PHANTOM™ Haptic Device are not grounded, hence the user feels the inertia of their stators when moving the device; this artifact is minimized by the mechanical design and the small size of the actuators. A different approach is necessary when stronger forces (hence bigger motors) are required. Force Dimension developed a new design for 3D devices which can provide strong forces while keeping all the three motors stationary; the resulting devices, called Delta and Omega are simple and strong, with a contained increase in perceived inertia with respect to the PHANTOM™ [39, 40].

Finally, the Ministick is a statically balanced device that combines low inertia and high resolution; it is based on three five-bar mechanisms which allows for stationary motors and direct drive transmission between the motors and the links [3]. The most recent implementation of this design is being used to investigate the perceptual properties of virtual textures [25, 136]; it produces a maximum continuous force of 1.4 N (5 N peak), which is sufficient for finger scale haptics, and has a nominal resolution of 9.6 μm according to the design paper [136].

A study on the detection of the orientation of sinusoidal gratings showed similar thresholds when the task is performed virtually with the ministick and with the bare finger on real surfaces; the authors interpreted these results as a validation of the rendering capabilities of the ministick [139]; this conclusion is not completely supported by the results because of two reasons: the different mechanisms involved in direct and indirect touch should be discussed and quantified. Second, according to the force constancy theory [24], changing the stiffness of the virtual wall supporting the texture could affect the thresholds because of the specific algorithm used in the experiment to convert the geometrical profile in force field.

2.2.2.7 Multi DOF Force-Torques

Both the Sensable PHANTOM™ and the Force Dimension Delta are available in variants that provide 6 DOF force/torque feedback, by adding encoders and motors at the tooltip of the device. These extensions increase the perceived mass of the device and do not offer the same performance of specific multi-DOF designs.

An example of commercially available 6 DOF device is the MPB Freedom 6S, which combines a parallel design and cable driven transmissions to achieve low inertia, wide bandwidth, and high positional resolution. The main advantage of this device with respect to the other 6 DOF commercial interfaces is quality of the force delivered to the handle.

To avoid perceivable artifacts arising from friction and backlash in the mechanisms of the haptic device, a design based on magnetic levitation can be used: with a workspace of ± 12 mm in translation and ± 7 deg rotation, the magnetic levitation haptic device developed by Berkelman is very limited in its applications, but it was successfully used to display virtual textures [142]. The resolution of the device (5–10 μm) and the bandwidth of 120 Hz (-3 dB) offer a level of performance adequate for the texture rendering; nevertheless, guaranteeing the passivity of the device could be problematic, since position quantization is known to introduce non passive behaviors in low friction devices.

2.2.2.8 Frequency Response and Artifacts

The aforementioned commercial haptic devices have been extensively used for haptic research both for developing rendering algorithms and for studying the perceptual response of the somatosensory system. However, the engineering limits of such interfaces are not yet fully understood, and the manufacturer specifications often lack essential information such as the physical bandwidth of the device, whose effect on haptic texture perception was explored by Wall and Harwin [144].

A critical study on the engineering characteristics of the PHANTOM™ haptic device was conducted by Cavusoglu et al. [13]; the most interesting aspect of their study, with respect to the topics discussed here, was the confirmation of frequency dependent amplification of the force signals, which can greatly distort the rendering of haptic textures. Further analysis of the frequency response of the PHANTOM™ was presented by Kuchenbecker et al. [81], where a high order invertible linear model of the device dynamics was proposed as a solution for shaping the open loop acceleration response of the device. This approach cannot be directly applied to closed loop haptic simulation, but represents an interesting development for time-based haptic synthesis of virtual textures, which is reviewed in Sect. 2.5.3 at page 33.

Contrary to the PHANTOM™, custom made devices are usually not investigated as extensively as it is required to guarantee artifact-free rendering. In particular, the ministick is an interesting device but little is known about its frequency related characteristics.

To ensure an artifact-free haptic signal, a haptic device is often analyzed in the context of control theory; this approach provides a theoretical framework which can be applied to different haptic algorithms.

2.3 Control Theory and Haptics

In general, two distinct notions are used to analyze the control properties of haptic systems: passivity and stability. Dynamic system, such as haptic devices, react to an input signal by changing their state and by providing an output signal. In the case of an impedance style force feedback device, the input is the motion of the handle by the user, and the output is the actuation (force/torque) applied to the handle.

A dynamic system is said to be stable if for every bounded input signal, it responds with a bounded output signal; simply put: if the user moves the handle with a finger, the device does not try to tear the limb apart from the owner. The same stable dynamic system exchanges energy with the environment, but no restriction is made on this exchange, as long as the output does not become unbounded.

The passivity of a system, on the contrary, is defined by constraining the energy flow between the system and the environment: a system is passive if when perturbed it responds by dissipating part of the input energy, thus returning less energy than it receives. More formally a one port system with effort $\mathbf{F}$ and flow $\dot{\mathbf{v}}$ is passive if:

$$\int_0^t \mathbf{F}(\tau)^\mathsf{T}\dot{\mathbf{x}}(\tau)\,d\tau + E(0) \geq 0 \tag{2.4}$$

for $t \geq 0$ and for every function pair $\mathbf{F}$, $\dot{\mathbf{x}}$. where $\mathbf{F}$ the force generated by a haptic device at the handle, $\dot{\mathbf{x}}(t)$ the velocity trajectory of the handle, and $E(0)$ the energy of the system at $t = 0$. In passivity literature, dissipated energy has positive sign, hence the flow is the velocity at the handle with negative sign. In other words, during a passive haptic interaction, the haptic device dissipates part of the energy of the user motion. While a passive system is clearly stable, stability does not imply passivity: for example a stable haptic device can generate bounded, periodic, self sustained oscillations around an operating point, which is a non passive behavior.

The use of passivity is widespread in the haptic community because it allows the decoupling between the analysis of the mechanical haptic system and the properties of the human user. Moreover, a passive system does not generate sustained vibrations at the tooltip, which is a typical artifact that spoils the haptic experience. The latest results in passivity analysis for haptic devices offer a set of constraints on the parameters of the virtual environments, which are based on the physical characteristics of the mechanical system (for example friction in the ball bearings and damping); most of the results refer to the simple case of the unidimensional damped linear virtual wall. The discussion of these results can be integrated with the limited attempts made for extending the passivity results to the non-linear and multidimensional case.

2.3.1 Passivity Results

The usual haptic device is controlled by a discrete sampled data controller, which in general reads the position of the handle, computes the force output, sends the appropriate signal to amplifiers which, in turn, deliver controlled power to the motors. This process is repeated as fast as possible, generating a periodic process. The frequency of this process is the sampling frequency, and during the interval between two iterations the forces are kept constant according to the zero-order-hold sampling scheme ($\mathcal{ZOH}$). Typically, artifacts emerge when implementing a haptic virtual environment: the delay between the acquisition of the position and the actuation of the motors and the sampling period introduce an unwanted energy imbalance, resulting in a series of oscillations and vibrations.

2.3.1.1 Physical Damping and Sampling Period

Colgate and colleagues studied the problem of rendering passive virtual walls of stiffness K_V and damping coefficient B_V; they identified the relationship that ensure passivity in:

$$\frac{K_V}{2T} + |B_V| \leq B_P, \tag{2.5}$$

where T is the sampling period and B_P is the mechanical damping in the haptic device [27]. This classical equation has been extended by Colgate et al. in [26] to account for the effects of a first order filter used to smooth the velocity signal:

$$\frac{K_V}{2T} + \frac{B_V T}{2\tau + T} \leq B_P \quad (B_V \geq 0), \tag{2.6}$$

$$\frac{K_V}{2T} - B_V \leq B_P \quad (B_V \leq 0) \tag{2.7}$$

which includes the time constant τ of the filter. In a following paper [28] the authors explore the relationship between passivity and stability, showing that passivity imposes more conservative constraints than stability on the virtual environment parameters; particularly, as the device damping B_P decreases, the range of parameters that allow for stable rendering is bigger than the range of passive environments.

Among different extensions of those equations to multidimensional virtual environments, a very interesting formulation can be found in [97] where Mahvash and colleagues found two sufficient conditions for passivity for a delayed and sampled system: let $a(t)$ be the acceleration of the haptic device, if

$$|a(t)| \leq \frac{\sigma_0(\mathbf{S}_P)}{T\sigma_0(\mathbf{B}_P)}, \tag{2.8}$$

where $\sigma_0(\mathbf{S}_P)$ is the minimum singular value of the physical friction matrix of the haptic device, and

$$|\mathbf{F}(t_k) - \mathbf{F}(t_{k-1})| \leq \frac{\sigma_0(\mathbf{B}_P)}{2T}|\mathbf{x}(t_k) - \mathbf{x}(t_{k-1})| \tag{2.9}$$

are verified, then the resulting haptic interaction is passive; due to the multidimensional nature of the system, the physical damping B_P is replaced by the minimum singular value of the damping matrix $\mathbf{B}_P$. For a linear virtual wall equations (2.5) and (2.9) are equivalent.

Another analysis of the passivity of non-linear virtual environment is provided in [102]; in this formulation, the non-linear environment is described in terms of the maximum differentials of the force with respect of position, velocity, and acceleration. Moreover, this paper also considers the influence of the virtual coupling, which is an approach to stability proposed by Adams and Hannaford [2].

2.3.1.2 Friction and Encoder Quantization

Two recent studies have shown that encoder quantization can be responsible for the non passive behavior of certain haptic simulations, particularly in devices with low inherent friction [1, 34]. In [1], Okamura and colleagues found a simple and elegant solution for the energy gains due to quantization when rendering a linear virtual wall:

$$K_V \leq \min\left(\frac{2B_P}{T}, \frac{2f_c}{\Delta}\right) \tag{2.10}$$

is an extension of Eq. (2.5) that accounts for friction (f_c) and encoder quantization (Δ). While this solution was found by analyzing the energy balance of the haptic simulation in the spatial domain, Diolaiti et al. [34] used the machinery of control theory to determine the behavior of a haptic device with friction and spatial quantization. This analysis confirmed that globally stable (passive) interaction is possible if Eq. (2.10) holds. Moreover, the authors identified three possible behaviors if the passivity equation is not met: limit cycles of very small amplitude (less than one encoder count) can happen if $K_V \leq \frac{2B_P}{T}$ and $K_V > \frac{2f_c}{\Delta}$; on the other hand, if the haptic device has enough friction but not enough damping, the simulation can be either locally stable or unstable depending on the initial velocity, which can be too high for the friction to fully dissipate the energy of the system, resulting in unstable rendering.

2.3.1.3 Time Based Passivity

A simple real time method to ensure passivity has been investigated by Hannaford and colleagues [53, 127]. The energy exchange between the virtual environment and the haptic device is monitored by the so-called Passivity Observer, and when the energy balance shows non passive behaviors a programmable virtual damper (called Passivity Controller) is activated to remove energy coming from the virtual environment. The Passivity Controller approach does not model either the device dynamics nor the users behaviors, hence its wide applicability; in some cases the unmodeled dynamics of the haptic device could introduce artifacts, and some heuristic corrections based on the virtual environment might be needed [128]. Finally, this time based approach can be used to control a flexible link, if a dynamic model is known [126].

2.3.1.4 Wave Haptics

Niemeyer and colleagues introduced the concept of wave variables in teleoperation [108, 109], and applied the same formalism to haptics [35]. This approach is based on a change of variables that converts the usual effort/flow pair into a so-called wave space, whose main property is the passivity under time delay. In practice, it is used to design an analog amplifier/controller which exploits the electric motor dynamics to improve the rendering of stiff passive virtual contacts. The results are encouraging in terms of high frequency response, but the difficulties of designing virtual environment in wave space as well as the requirements for analog controllers are limiting the adoption of this approach.

2.3.2 Stability of Haptic Systems

As already mentioned, stability is a less restrictive condition than passivity, but it requires a model of the human operator, which is reflected in a series of assumptions on the mechanical properties of the human hand or arm as well as on the behavior of the user. Sometimes, stability is studied in the worst case scenario, which happens when the user does not interact with the device [34].

Nevertheless, multiple stability conditions have been proposed based on different models used for describing the user/haptic device system. Under the hypothesis that both the human arm and the device can be modeled as second order systems, Minsky and colleagues found that, for a virtual wall to be stable, the inequality

$$\frac{K_{\mathrm{T}}T}{2} \leq B_{\mathrm{T}} \tag{2.11}$$

must hold, where T is the sampling period in seconds [105]. For purely damped virtual environments, equation

$$\frac{T}{2M_{\mathrm{T}}} \leq B_{\mathrm{T}} \tag{2.12}$$

must be verified, K_{T} being the total stiffness, B_{T} the total damping, and M_{T} the total effective mass of the system haptic device—human arm.

Hayward and Bonneton extended the analysis to a system with a delay of a single time sample T_{s} and found that

$$B_{\mathrm{V}} \leq B_{\mathrm{P}} + \frac{4M_{\mathrm{P}}}{3T_{\mathrm{s}}}, \tag{2.13}$$

$$K_{\mathrm{V}} \leq \frac{2B_{\mathrm{P}} + B_{\mathrm{V}}}{3T_{\mathrm{s}}} \tag{2.14}$$

must hold for virtual wall to be passive, where K_{V} and B_{V} are the parameters of the virtual wall being rendered, and M_{P} and B_{P} are the physical parameters of the device [10].

Later, Gillespie and Cutkosky studied the problem of simulating a ball bouncing on a virtual surface: for this scenario they developed a new controller, based on dead-beat control, which solves the problem of the asynchrony between the digital sampling and the virtual contacts, because the ball crosses the threshold of the wall between two samples. This controller could then be used to render virtual walls, once a model of the user, and of the device, is provided [46].

The range of stable stiffnesses renderable by a device is limited; to improve the perception of hard contact and the stability of the system Salcudean and Vlaar proposed to add a pulse of braking force upon contact with the virtual wall [129]. The braking force is shown to partially compensate the deterioration of performance due to discretization.

The stability boundaries of a haptic device (modeled as a damped mass) rendering a damped virtual wall were explored in [66]; the authors found that the maximum stable virtual damping renderable depends on the physical damping in the device, and it is independent from sampling period and time delay. The authors also discuss the maximum stable stiffness and compare it with the passivity condition in Eq. (2.5): if there is delay in the rendering, Colgate's equation is not valid, because the range of passive parameters (K_{V} and B_{V}) derived from such equation contains values outside the stability boundary of the delayed system; this result implies that the passivity condition is a best case scenario for the virtual environments. Similarly, the influence of delay and damping on the stability boundaries has been studied and experimentally validated in a follow up work [45].

2.3.3 Virtual Coupling

The unification of the impedance and admittance approach to haptics is explored in a paper by Adams and Hannaford [2]. The authors propose that admittance style displays could be used to display impedance type virtual environments and vice versa, for a total of four different combinations of devices (admittance/impedance) and virtual environments (admittance/impedance). The stability of those four kinds of systems is analyzed and passivity can be enforced by using a virtual coupling element, which acts as a converter of the force/velocity quantities between the virtual environment and the device. For impedance/impedance systems, the maximum stiffness of the elements in the virtual environment must be limited; the role of the virtual coupling in this case is to limit the stiffnesses that the device transfers to the user, so that if a virtual element is too stiff, the system does not generate oscillatory artifacts; however, when interacting with stiff environments the user would feel the compliance of the coupling and not the stiffness of the elements: this trade-off between stability/passivity and transparency must be taken into account when designing the virtual coupling. In conclusion, the virtual coupling simplifies the design of both the virtual environment and the device by decoupling their stability/passivity properties.

This brief review of the major results of control theory applied to haptics, shows the complexity of the problem of rendering stable or passive virtual environment. The general reliance of stability analysis on a model of the human user is definitely the major drawback; for this reason the passivity analysis is often preferred.

2.4 Texture Perception

The perceptual mechanisms underlying the tactile exploration of materials are still under investigation; it is still unclear which characteristics of the objects, and specifically of the textures, are perceptually more relevant. The link between physical measures of an object and their perceptual influence has been elusive so far; a recent work by Tiest et al. [141] tried to correlate the perceptual similarity of 124 samples of common objects with the physical roughness of their surface and the compressibility. By using multidimensional scaling (MDS), they established that subjects sorted the 124 objects according to four psychophysical dimensions, none of which coincided with either physical roughness or compressibility; moreover, these two properties are mapped in an horseshoe shaped curve in MDS space, which led to the conclusion that the perceptual space is not Euclidean. Previous studies conducted with MDS led to different conclusions: the perceptual spaced looked Euclidean and the dimensionality was between 3 and 4 [61, 63, 120]; these discrepancies can be explained by the limited sample size used in those early experiments, at most 25.

To label the dimensions, subjects are in general required to describe the stimuli according to a series of predetermined adjectives. There is consensus on the nature of the most important directions: stiffness/compliance (on the continuum soft-hard) and roughness (smooth-rough) are the most important characteristics, but their influence is often combined in the first MDS dimension, thus suggesting that they are not independently used to perceptually characterize surfaces.

Picard et al. [120] conducted a free-sorting experiment on 24 car seat cover samples; the most relevant conclusion regards the issue of semantic differences between languages, because the most important dimension used by the subjects was labeled as soft/harsh, and roughness did not play an independent role. This results might be explained by the very specific set of stimuli as well as by the different nuances that the word roughness carries when applied to different materials.

Hollins et al. investigated the individual variations of this perceptual tactile space and found that while some subjects used mostly those two dimensions to group objects (rough/smooth and hard/soft), others discriminated also based on a third direction, labeled stickiness [63]. This result is very significant because it shows that, when a limited set of stimuli is presented, stickiness (hence friction) is not a third independent dimension for all subjects. In their previous study they found again a three dimensional space, where the first dimension was clearly aligned with roughness and the second with stiffness [61].

In conclusion, the space of tactile perception has not yet been fully understood, but it is clear from the literature that: at least four dimensions are involved, the space is probably not Euclidean, and that, although coupled with other properties, roughness has a strong representation in this space.

From the MDS analysis, it appears that roughness/softness is the dimension that is most related to texture perception; unsurprisingly then, roughness has been the topics of truly many studies, which approached the investigation of roughness perception from two different points of view: classical psychophysics experiments, based on subjects reporting on texture properties, explained which elements of the haptic interaction with surfaces influence the percept of roughness. Among those elements are: the mode of touch (active or passive, direct or mediated by a probe), the geometric properties of the surface (microstructure, macrostructure), the material properties of the texture (friction, compliance, temperature), and exploratory conditions (speed of exploration).

The second approach is based on the identification of the physiological events that determine the perception of roughness. Typical experiments consist of recordings of neural activity of monkeys in response to tactile stimulation. Under the assumption that neural mechanisms of monkey and humans are correlated it is possible to infer neural code for roughness by performing a psychophysical experiment on humans, and then record the neural response of monkeys exposed to the same stimuli. For example, human subjects could be asked to estimate the roughness of some gratings, then the same surfaces are applied to the fingerpad of monkey and the neural response is recorded (either centrally in the somatosensory cortex of the brain or peripherally in the median nerve that runs in the wrist); for an example see [138].

2.4.1 Early Work on Roughness of Textures

The first qualitative results regarding the perception of rough textures were collected by David Katz (1925), who tried to isolate the effect of a number of exploratory conditions [151]. With regard to this book, the most important result is that roughness perception requires lateral motion between the skin (or the probe) and the surface under investigation. Moreover, rough textures can be reliably perceived through a probe: subjects can discriminate 14 different samples of paper (from smooth waxed paper to coarse cloth paper) by stroking them with a pencil. This ability is purely mediated by vibration, because once the pencil is covered in a vibration damping cloth, the discrimination performance is greatly impaired. These two observations confirm that roughness can be conveyed to a user through a haptic device. Among the conditions tested, Katz noticed that changes in exploration speeds have minimal influence on roughness when using a bare finger, but when speed exceeds 60 cm/s, roughness disappears altogether and a pain sensation arises. In similar experiments, Zigler noticed that roughness does not arise from stationary contact between the finger and a surface, but it requires a relative movement between them [100]; the author also reports that the experience of smoothness is mostly elicited when the interaction force is moderate or light.

More recently, Stevens and Harris showed that subjective estimates of roughness and smoothness are linear functions of the grit number of emery cloths when plotted in log-log space (the grit is a measure of inter element spacing of the particles on the surface of the cloths) [137]. Moreover, the authors proved that estimates of roughness and smoothness (across subjects) are perceptually reciprocal, both when the estimation is absolute or relative to a standard cloth. A similar study by Ekman and colleagues proved that roughness/smoothness are reciprocal also intra-subject, although the psychophysic functions differ substantially between subjects [38].

Finally, in contrast with the common assumption that active touch is generally better than passive touch [44], Lederman showed that roughness perception with a bare finger is not significantly affected by the mode of touch [87] an only marginally when mode of touch is combined with the exploration speed [88]. The kinesthetic perception and the proprioceptive sense, then, contribute marginally to roughness perception, and their role has been often overlooked in studies related to texture perception. Also, due to this marginal contribution of the proprioceptive sense, data obtained from experiments performed under active and passive conditions can be directly compared, when they deal with roughness.

2.4.2 Bare Finger—Macro-textures

Although this topic had been studied previously, Lederman's PhD work is arguably the first systematic attempt to understand the psychophysics of roughness; by drawing inspiration on Stevens' work [137], the author based most of her experiments on roughness estimation of different kinds of surfaces: linear metal gratings, raised dot patterns, and sandpaper among them. This research has been extended to virtual environments generated by a force feedback joystick [104], a haptic mouse [74], and a PHANTOM™ [92].

Lederman conducted a set of experiments with linear gratings, produced by cutting parallel linear groves of different width on the surface. This process results in a series of raised ridges; these stimuli belong to the category of macro textures (with inter-element spacing >200 μm), and their estimated roughness was found to be significantly affected by a number of parameters, both when explored with a bare finger or through a probe.

When the surface is explored with a bare finger, the most important effect is usually the groove width both in metal [85, 88, 89] and plastic surfaces [83]. The psychometric function relating groove width and estimate of roughness has an inverted U shape in log-log space with a peak above 3.5 mm, confirming the results of Connor et al. who showed that roughness peaks for embossed dots spaced 3.2 mm [30]. A decrease in perceived roughness occurs as the ridge width increases, but this significant effect is not as large as the groove width influence [85, 89], similar effects of groove and ridge size were found by Sathian et al. [130]. An interesting comment on the metal gratings used by Lederman can be found in [85], where the authors reported that two different manufacturing processes created stimuli that exhibited a different finish due to the microstructure; the stimuli felt "somewhat different" although their macro structure was similar. This effect can be explained by the sensitivity of the finger to microtextures; the features left on the surface by the manufacturing process were smaller than 200 μm but contributed to the overall roughness perception.

The U shaped relationship between inter-element spacing and roughness is disputed by Chapman [101], Smith [134], and colleagues who used non dithered plastic raised cones (approx 1 mm in diameter) and found that roughness estimates were monotonic log-log functions of the inter-element spacing (from 1.5 to 8.5 mm). The most probable cause of the quadratic relationship found by Lederman, Connor, and others is the small elevation of the textural elements over the surface (grooves or dots), which causes the finger to touch the bottom of the groove, when the width exceeds 3 mm. This contact between the finger and the flat surface was less likely to happen in Chapman and Smith's case because the patterns had taller textural features.

Focusing on the interaction forces between the finger and the surfaces, Lederman found that the normal force applied by the fingertip on the surface [89] has a large effect on roughness: more pressure equals rougher percept. Conversely, shear forces applied to the fingertip impair the estimates of roughness when subjects explore surfaces through layers of paper [86]; these experiments extend the findings of Gordon and Cooper, who found that asperities of surfaces are better detected when a piece of paper is interposed between the surface and the skin and moved together with the finger [51]. The explanation of this phenomenon can be found in the work of Smith and colleagues [134], who proved that roughness correlates with the rate of change in lateral force when subjects explore plastic surfaces with raised dots. As a consequence, constant shear forces could mask lateral force variations, and reduce the perception of roughness.

Furthermore, Smith showed that lubricating the surfaces (hence changing the friction coefficient) does affect the estimate of roughness because it reduces the lateral force variations. This conclusion is consistent with Ekman [38] who measured a log-log linear friction to roughness psychometric function. In contrast with those results, Taylor and Lederman showed a negligible effect of friction when exploring metal gratings [140]. Since the friction coefficient was computed when the finger stroked *along* the ridges, they did not consider the effect of the variations due to geometry; this effect was instead included in the mean friction computed in the other two studies (Smith's and Ekman's).

2.4.2.1 Temporal Determinants

The effects of spatial features, and particularly the spacing between textural elements, have been confirmed; however, there is a debate over the influence of temporal determinants on the perception of textures. Lederman's experiments showed that exploration speed does not influence the roughness perception of gratings [85, 88], even when the pitch of the surface is varied. From this observation it was concluded that the frequency of vibration of the skin does not affect roughness estimation. The same speed invariant behavior was observed by Chapman and colleagues for raised dots patterns, they also confirmed that frequency is not a determinant for macrotextures and mentioned that mode of touch was not significant [101]. To completely exclude the influence of frequency on roughness estimation, Lederman and colleagues used selective frequency adaptation to impair the perception of vibrations and asked the adapted subjects to estimate roughness. Two groups of adapted subjects, 20 and 250 Hz, was confronted with a control group and, although adaptation was measured, no significant effect on the estimates of roughness was found [90]. Hollins and colleagues confirmed, about 20 years later, that frequency adaptation does not affect the judgment of roughness for macrotextures, but greatly impairs the perception of fine textures [64]. Discordant results are presented by Cascio et al.; they found a significant influence of frequency on roughness perception when estimating and discriminating macro-gratings with various ridge widths, but no such effect could be found for stimuli of different groove width [12].

2.4.2.2 Spatial Intensive Model for Macrotexture Perception

The findings of the perceptual works on gratings can be explained "by the spatial—rather than temporal—aspects of their biophysical interaction with the hand [59]". Taylor and Lederman [140] introduced a model to interpret the spatial determinants of roughness: they propose that roughness increases as a function of the mean volume of skin displaced by the mechanical interaction with the surface. A notable omission in Lederman work is the study of texture perception and discrimination in the case of no movement between the finger and the surface. According to Katz, there cannot be any roughness estimate without movement; Hollins and colleagues investigated this issue quantitatively and proved that movement increases the discrimination of fine sandpaper (particle size < 100 μm), but it has no effect for coarser samples. This finding is the basis for the so called duplex theory of texture perception: Katz already proposed that tactile perception is mediated by two different channels, one sensitive to pressure and the other related to vibrations. Hollins and colleagues confirmed that, at least for texture perception, the duality theory holds and has a neurophysiological explanation [60].

2.4.3 *The Duplex Theory of Texture Perception*

Building on Lederman's spatial model, Hollins and colleagues proved that perception of fine textures is mediated by the perception of vibration, while macro textures are mostly perceived by sensing their spatial features. In experiments conducted with sandpaper, the transition between spatial and temporal perception happened for element sized of 100 μm [60], whose spatial period is around 200 μm. To further confirm this finding, experiments with sandpaper showed that textures with spatial period close to this limit feel rougher if an external vibration is applied to them [62]; because this increase in roughness is not dependent on the frequency, it is very likely that finer surfaces feel rougher due to the amplitude of the vibrotactile stimulation. Finally, as further confirmation, adaptation to frequency impairs the perception of finer surfaces but not coarse ones [64].

2.4.4 *Neurophysiology of Texture Perception*

Validation for the duplex theory is provided by the vast body of literature on the neurophysiology of touch in monkeys, whose somatosensory system closely resembles the human's.

The tactile sensation arises in the skin, when mechanical deformation of the tissue is converted in neural impulses sent to the brain. This neuro-mechanical interfaces in the skin are called mechanoreceptors, and in the human glabrous skin there are four different kinds of mechanoreceptors, which can be identified both by the anatomy and by the characteristics of the neural response they generate. For the purpose of this book, only the most salient characteristics of the mechanoreceptors, their functions, and their link to virtual textures are reviewed. For in depth reviews on mechanoreceptors please consult [70, 72]. Of the four mechanoreceptors in the human skin, two are innervated by slow adapting fibers (SA1 and SA2) while the other two are rapidly adapting (RA1 and RA2).

2.4.4.1 SA1 Afferents and Receptors in Human Skin

At the interface with the dermis, the epidermis is structured in ridges, whose tips are innervated by branching myelinated neural fibers that terminate in the Merkel disks; in particular, Merkel endings cover 80% of the epidermal ridges area in monkeys fingerpad [117]. Two organizations of Merkel endings were found: branching afferents innervated either clusters (70 μm in diameter) or chains of Merkel endings with length greater than 200 μm. These fibers are called SA1, and are believed to respond to the mechanical strain energy density in the skin [72]. These receptors and afferents are found both in human and monkeys. When the skin is indented, SA1 fibers respond with a rate of impulses that increases with the indentation amplitude, but there is debate whether this relationship is linear (e.g., [8] on monkeys) or sublinear ([76] on humans). The SA1s have been shown to be the most reliable receptors that respond according to the spatial properties of the object being touched: curvature (human [50] and monkey [48]), edges (human [119], monkey [8]), orientation of curved surfaces [36], and, most importantly, roughness [29, 30, 71, 130].

The response of single SA1 afferents to textured surfaces could not be correlated to roughness estimates of humans, but, as a population, SA1 receptors do represent roughness, the code being the spatial variations of firing rate in the population of SA1 afferents: this neural code can be clearly predicted by filtering the textured surface with a spatial Gabor filter [29]. Moreover, the somatosensory cortex of monkeys, SI—area 3b where SA1 afferents are projected in the brain, contains neurons whose receptive field structure suggests the function of enhancing local features of a surface (like edges) [33], which could further explain the SA1 code for roughness. This theory about texture perception is disputed by Chapman and colleagues, who propose a simpler rate-based neural code which is then integrated with motion information in the somatosensory cortex [14].

Sathian and colleagues showed that SA1 response was unreliable indicator of roughness of gratings with small groove width [130], in accord with the large size of their receptive field [143]. Studies on gratings moving sinusoidally indicate that of the 2 spatial and 2 temporal determinants, only one spatial variable (groove width) and one temporal variable (peak temporal frequency of the grating spatial cycles) affect the neural response independently. The other two determinants (ridge width and peak speed of movement of the gratings) affect the neural response only in the measure that they change the groove width or the peak frequency [47]. Finally, the detection of gratings is improved when the features are presented *parallel* to the dermal ridges on the skin, with respect to *orthogonal*; recordings in monkey show that SA1 population responds more vigorously to gratings parallel to the ridges, matching the results on human performance [148]. This phenomenon could be explained by the ridge either acting like a lever, thus magnifying the strain at the location of the Merkel disks, or, by the same mechanical property, ridge movement could recruit more Merkel disks, for the same skin deformation.

Interestingly, SA1 receptors responds to forces aligned along preferred directions, thus encoding the directionality of the force stimulus [7]; this property is however shared with the SA2 and RA1 sensory systems.

This brief list of findings seems to indicate that SA1 afferents population is responsible for the detection of shape [49, 50] and mediate the sensation of roughness, for medium sized feature. The purely spatial hypothesis on texture perception, however, conflicts with findings from [12], as mentioned above; this discrepancy could be explained by the higher sensitivity of SA1 to dynamic stimuli [70]; however, the poor performance of SA1 on fine textures confirms the role of dynamic mechanoreceptors in the perception of textures (innervated by RA1 and RA2 fibers), thus validating the duality theory.

2.4.4.2 RA1 Afferents and the Meissner Corpuscle

About 150 RA1 afferents per cm^2 innervate the human skin and terminate in Meissner corpuscles, which are a fluid filled structure lodged between the adhesive ridges in the skin (dermal papillae), closer to the skin surface than the Merkel disks. For a description of the Meissner corpuscle please refer to [116].

The proximity to the surface of contact is responsible for the low thresholds needed to stimulate the afferents; nevertheless, the RA1 afferents have very poor spatial discrimination because they respond with uniformity to stimulations over the entire receptive field (3–5 mm in diameter) [70]. A physiological study by Paré and colleagues established a density of approximately 45–60 Meissner corpuscles per mm^2 of monkey skin; moreover the innervation patterns are complex [117]: "Finally, a given MC may be innervated by one axon that supplies one combination of multiple MCs and another axon that innervates another, albeit partially overlapping combination. This anatomical arrangement suggests that the MCs are part of a partially shifted, overlapping continuum of afferents that include a mix of differing resolutions." The dense innervation implies that each innerving fiber branches multiple times to reach different Meissner corpuscles.

The most important function ascribed to the RA1 system is the detection of slip during precision grip and manipulation of tools [135]; in particular, it is the only system that detects slip for surfaces with features height in the range 2–8 μm. For a review on the importance of cutaneous receptors on grip forces please consult [149].

The role of RA1 afferents on roughness perception is not yet clear [59]; the response of RA1 to textures moving on the fingerpad is greatly impaired by the limited spatial resolution, but due to their very low thresholds, RA1 afferents are responsible for the detection of very small features on smooth surfaces, for example RA1 can detect dots as small as 550 μm in radius and 2 μm in height; in comparison, the SA1 are four time less sensitive to dot height [82].

2.4.4.3 SA2 Afferents and Ruffini's Corpuscle

The neurophysiological similarities between the hands of humans and monkeys are striking; however, the SA2 system is absent in monkeys [70]. Whereas the response of SA2 to lateral skin stretch is established, there is strong debate on the nature of the mechanoreceptor responsible for the sensation. Commonly thought to innervate the Ruffini corpuscle, the SA2 afferents account for 15% of the primary afferent fibers of the median nerve innervating the human hand; however, immunofluorescence of the mechanoreceptors and innervation of the monkey finger, failed to isolate a single Ruffini corpuscle [117]. Further studies on human tissue isolated only one candidate Ruffini mechanoreceptor in the finger and concluded that "...it seems unlikely that Ruffini corpuscles in human glabrous skin account for all but a small proportion of physiologically identified SAII afferents [118]."

Despite the uncertainty over the origin of the mechano-neural transduction, two functions of the SA2 system are known: the detection of pre-motion deformation of the skin against an object; and, most importantly, the perception of the hand position through the pattern of strain on the skin of the fingers (integrated with kinesthetic information from muscle spindles and joint afferents).

By its nature however, the SA2 system seems to be marginally involved in texture perception, and it has not been investigated in this context.

2.4.4.4 RA2 Afferents and the Pacinian Corpuscle

The RA2 system then presents very high sensitivity and responds to a relatively narrow range of frequencies (100–300 Hz). The mechanoreceptor responsible for the transduction of vibration into neural signal is the Pacinian corpuscle. Lodged deeply in the dermis, this ovoidal lamellar structure (varying in size between 1 and 4 mm in adult humans) is exquisitely sensitive to vibrations (in the order of 10 nm and more in amplitude, with a mean of 40 nm) and has a receptive field that is not as spatially defined as RA1 and SA1 [72]. There are approximately 350 Pacinians in each finger and 800 in the palm, each usually innervated by a single different fiber. In monkeys, approximately 2% of corpuscles are innervated by a branching axon [117].

The response of RA1 and RA2 afferents to vibrotactile stimulation is presented in a seminal paper by Mountcastle and colleagues, who studied the response of the skin to vibrations applied by probes with spherical tips (diameter 0.5 to 3 mm) [138]. The detection thresholds they measured are clearly bimodal, when plotted as a function of the stimulation frequency: further study on monkeys showed that RA1 fibers response peaks at about 30 Hz, while RA2 are extremely sensitive to stimulations at 250 Hz. The amplitude detection thresholds at 30 Hz are one order of magnitude bigger than at 250 Hz. This difference in sensitivity can partially be explained by the mechanical properties of the skin: mechanical wave conduction is very poor at 30 Hz, in fact vibratory stimulation must be within 5 mm from the center of the receptive field to excite the relative RA1 afferents. Notably, subjects can rely on RA1 to distinguish biharmonic signals differing only in phase (30 Hz + 10 Hz sinusoidal waves) while the Pacinian afferents cannot (100 Hz + 300 Hz sinusoidal waves) [5]. To explain this finding, an intensive model for pacinian representation of vibrations was proposed: the hypothesis was that vibrations could be distinguished according only to their pacinian weighted response. This model was partially rejected because the discriminability of multi-harmonic vibrations depended also on temporal properties [6].

It is established that roughness perception and texture detection of fine textures depends on the 250 Hz Pacinian vibrotactile channel: numerous studies by Hollins and colleagues established the role of the Pacinians by showing, among others, that adaptation to 250 Hz impairs the perception of fine textures, while coarser textures are not affected [64]. In following studies, by measuring the vibrations of the skin when scanning a surface, they concluded that "...the roughness of a fine surface (spatial period < 200 μm) is a function of the Pacinian weighted power of the vibrations it elicits."

Pacinian adaptation has also been studied with MDS: a three dimensional space was found for the perception of raised-dot textures (14 conditions $1, 2 \ldots, 7$ mm inter-element spacing, adapted and non adapted) [43]; adaptation to a 250 Hz vibration did not impair the perception of roughness of the surface, but made the individual dots feels smoother, which is consistent with the duality theory, since macro-textures are not perceived via the Pacinian system.

2.4.5 From Bare Finger to Probe

In a more recent study, the effect of adaptation to textures was studied in both direct and indirect touch. Exposure to a 416 μm texture proved to significantly reduce roughness estimates in indirect touch, as well as in direct touch for fine textures, but for coarse textures and direct touch no such effect was found [65]. This result is consistent with the intuition that, in indirect touch, only temporal determinants are available for texture perception.

The ability to distinguish surfaces and experience roughness through a probe was already described by Katz, and the MDS analysis of texture perception confirms that roughness is a perceptually relevant dimension both in direct and indirect touch [150]; this study also links roughness with the power of the vibrations in the probe during contact with a texture.

The estimated roughness of plates with dithered raised dots through a pen-like probe presents a log-log U shaped relationship with the inter-element spacing, similarly to what is observed for bare finger [73]. The inversion point of this function depends on the size of the tip of the probe: the smaller the probe the lower the inversion point, also the larger the probe the smoother the surface [84]. When a plastic sheath is used to remove the spatial component of stimulation to the fingerpad, the roughness discrimination is greatly impaired; on the contrary, the effect of latex glove (both on bare and sheathed fingers) is negligible, because the glove changes only the coefficient of friction at the contact site [73]. In both passive and active touch with a pen-like probe the largest effect on roughness is ascribed to inter-element spacing, but speed is indeed significant, showing a much larger effect in passive than in active condition [91], and the inversion point of the roughness function increases with speed; the most probable explanation for the speed effect is that when the subject (or the plate) moves faster, the probe sinks less in the texture, thus generating a smaller resistance; on the other hand when the spacings is sufficient for the probe to touch the bottom of the plate between dots, an inversion in the roughness estimate is expected. Interestingly, the larger the range of speeds used in the exploration the smaller the effect of speed on roughness estimation. Finally, as for direct touch, a higher normal force results in larger roughness estimates [84].

To account for these results Klatzky and colleagues proposed a preliminary probe/plate interaction model that predicts the drop point as a function of the probe size and plate geometry; the experiments, however, showed that the geometric drop point (of the probe in the texture) differs from the psychophysical inversion point, which then depends on dynamic parameters which are not captured by simply the drop point [75]. Finally in a summary paper, Lawrence and colleagues reiterated that roughness and groove width do not show a monotonic relationship, thus limiting the range of gratings that can be discriminated based on roughness only [83].

2.4.6 Texture Detection and Discrimination

Roughness is not the only perceptual quality that can be used for textures discrimination; in fact, it was found that the amplitude threshold of detection of a single

Gaussian bump of width $2 \times \sigma$ is linear in log-log space (exponent 1.3) with respect to σ. For this experiment, active touch with the bare finger was used, and the results are valid over a large range of bumps amplitudes ($0.15 \leq \sigma \leq 240$ mm) [95]. The absolute threshold for $\sigma \leq 1$ mm is close to 1 μm, and the critical parameter for detection is thought to be the maximum slope of the bump. This last observation is consistent with [134]. This study provides a relationship between a single textural element width and the height necessary for its detection; extending these findings on sinusoidal gratings, Nefs and colleagues determined that subjects could discriminate between textures that differed as little as 2 μm in height or 30 μm in pitch [107]. Previously, Morley et al. found a discrimination threshold of 45 μm for textures with different pitch under active exploration, which increased to 102 μm when the finger was kept stationary [106]; taking into account the difference in stimuli, these two studies substantially agree on the performance of the tactual perception of gratings.

The amplitude discrimination threshold reported [107] is smaller than the resolution of most haptic devices used for texture rendering. In conclusion, haptics devices operate somehow above threshold, because they cannot generate texture profiles with amplitude differences of less than 2 μm. Moreover, force feedback devices constrain the user motion on the virtual textures by applying a penalty force as the user crosses the virtual boundary; for example, a penetration of 2 μm in a texture with stiffness 10 N/mm would generate a force of 20 mN, which is approximately 2 grams/weight, fairly hard to detect when the inertia of the devices is at least 10 times bigger.

To conclude, the discrimination capabilities of the somatosensory system are a challenge not yet matched by haptic devices; as a consequence, the properties of the haptic devices must be taken into great consideration when designing psychophysical experiments and when analyzing the results.

2.5 Virtual Textures

As previously mentioned, the perceptual investigation of haptic textures can be traced back at least to the beginning of the 20th century [151]; conversely, the literature about the rendering virtual texture is much more recent, staring around 1990.

2.5.1 Geometry Based Methods

Margaret Minsky supervised the development of Sandpaper, the first system capable of rendering haptic textures [103, 105]. Her setup was based on a 2D force feedback joystick, which could display different texture patches. The device rendered only lateral forces, which, in some experiments, were proportional to the spatial gradient of the virtual height field $h(u, v)$ at the location of the virtual probe (x_1, x_2):

$$\mathbf{F}(x_1, x_2) = K_V \left[\partial h/\partial x_1 \quad \partial h/\partial x_2 \right], \tag{2.15}$$

where the height field is mapped according to $(u, v) = (x_1, x_2)$. The results confirmed that it is possible to convey the perception of a textured surface with a 2D joystick, hence 2D lateral force field can be effectively used to mimic surfaces with 3D micro structures. Furthermore, Minsky and Lederman demonstrated that the Sandpaper system was able to communicate to the user some form of roughness of the texture patches [104]. In this experiment, the perceived roughness could be almost entirely predicted by the maximum lateral force exerted on the hand exploring the texture.

The 3D extension of this method is explored by Hardwick et al. in [54]. Given a 3D coordinate frame (x_1, x_2, x_3) a flat virtual surface at $x_3 = 0$ and a 2D texture height field $h(x, y)$ mapped according to $x = x_1$ and $y = x_2$ the force is computed as:

$$\mathbf{F} = \left[-(\partial h/\partial x)\, R^z \quad -(\partial h/\partial y)\, R^z \quad R^z \right], \tag{2.16}$$

where

$$R^z = -K \min\left(x_3 - h(x_1, x_2), d^z_{\max} \right). \tag{2.17}$$

The value of $d^z_{\max}$ was set to approximately 0.1 mm to avoid instabilities. The device used in this investigation was a Immersion Impulse Engine 3000 with a maximum force of 9 N.

A different approach was proposed by Thomas Massie in his M.Sc. thesis [98], where he described some rendering techniques applied to the novel PHANTOM™ Haptic device. Massie's algorithm generates textures by perturbing flat surfaces in 3D with sinusoidal waves; the force arising from the interaction of the user with such textures is always normal to the flat surface and depends on the penetration of the virtual probe into the textures. This process results in convincing textures, which feel frictionless, due to the absence of forces tangential to the textured surface. For this same reason, Massie's approach remains very simple to implement and less prone to control instability. However, the virtual environment simulates a non physical interaction, which can introduce significant artifacts.

From these two pioneering works, it can be concluded that both normal forces and tangential forces can convey the feeling of textured surfaces. Basdogan et al. [4] combined the two stimuli by applying to haptics textures the bump mapping technique proposed by Blinn for computer graphics [9]. The feeling of roughness was generated by the perturbation of the normal to the flat surface with the gradient of the texture height field $\nabla h(u, v)$, with respect to the surface parameterization (u, v). The resulting force field has components both normal and tangential to the virtual untextured surface. This method was implemented on a PHANTOM™, and the users reported instability in the system; to solve this problem, Ho et al. [57] fine tuned the normal mapping model to limit the lateral force magnitude. Their solution is empirical but effective; however, in this book, a theoretical model of lateral force instability is proposed.

Since both 2D and 3D haptic displays could provide sufficient textural information, Janet Weisenberger et al. tried to determine which of the two modalities conveyed the most effective texture perception, [146]; the devices under investigation were a PHANTOM™ and an Impulse Engine 2000. On these devices they implemented both Massie's normal force algorithm and a viscosity based texture. The subjects were presented with virtual gratings aligned either along the X axis or the Y axis of the devices; their task was to discriminate the orientation of the virtual gratings. The strongest conclusion of this paper is the confirmation that both 2D and 3D devices can be used to render textures; although the Impulse Engine 2000 showed better thresholds than the PHANTOM™, no direct comparison between algorithms and devices was possible [145]. In a previous work they also showed that thimbles or pen like end effectors have no significant contribution on the discrimination of virtual texture orientation. Interestingly, Weisenberger mentions that renderings of sinusoidal surfaces with the PHANTOM™ exhibited instability when amplitude increased over a threshold; this phenomenon suggests that also Massie's algorithm for normal force fields suffers from control instabilities.

Similar considerations need to be made when dealing with 3D interfaces because they can render both 2D textures, laterally varying force fields applied to a flat surface, and 3D textures, which also have a geometric displacement along the normal to the same surface. Ho et al. discussed the perceptual equivalence between 2D and 3D square waves on a custom made 3D display. They concluded that, for small amplitudes and spatial periods ($<$ 1.5 mm), the two kind of textures are indistinguishable [58]. This result represents the first attempt to quantify the equivalence between two different texturing algorithms.

Another approach for texture synthesis is to draw inspiration from the roughness perception: the geometric model proposed by Lederman and colleagues [75] can be used to generate virtual textures. In the implementation by Unger and colleagues, the interaction between a spherical virtual probe an a surface with raised dots is simulated by constraining the sphere to be outside the surface; the roughness estimation of those virtual surfaces is different from the real ones, even when a velocity correction term is added [142]. No recordings of the user interaction are provided, making it difficult to assess if these results are due to the model, the device, or artifacts. Moreover, no assessment on the fidelity of the rendering can be made.

With a similar motivation, Otaduy and Lin extended the model by Minsky [104] and Hardwick [54] to account for the probe size [114]: in their algorithm, the normal penetration is "the vertical translation required to separate the probe from the textured surface". The force field is then function of the gradient of this normal penetration. Simulation show a good qualitative matching between the roughness magnitudes estimated in [91] and the maximum acceleration of the hand that this new model would generate.

Finally, a new algorithm for rendering 2D haptic textures was proposed by Hayward et al. [56]; used to render 2D textures on 2D haptic devices, it is based on the idea of directional finite differentiation: it renders a force proportional to the directional change in the virtual height field of a texture, but the force direction is always along the user motion. In this way, the virtual environment generates only working forces that either resist or facilitate user motion. This method, inspired by Minksy's and Costa's gradient techniques, can render discontinuous and non differentiable height fields without modifications, while gradient and normal based techniques require, at least, a differentiable height field.

2.5.2 Vibration and Reality Based Methods

Vibration is an important element of texture perception, its application to force feedback was investigated by Kontarinis et al. [78]; they showed that providing vibration feedback in a teleoperation setup definitely improved the user performance. Following this path, Okamura et al. proposed to augment the rendering of virtual object with data recorded during the exploration of real surfaces, which contained significant vibration components [110]. A stylus was stroked and tapped on real materials, and the vibration signal was collected by an accelerometer mounted on the probe. The resulting data was analyzed and correlated with factors such as velocity of exploration, normal force, and geometry of the surface, resulting in simplified deterministic models for generating virtual vibrations. For tapping, decaying exponential models were used. These open loop models, combined with standard force feedback techniques, were implemented on an Impulse Engine 2000 joystick, through which users were able to discriminate the properties of the original materials by feeling their virtual counterparts.

Okamura et al. extended this concept of reality based haptic rendering to a 3D device [111–113], and also developed a friction acquisition device that could be used to identify parameters for virtual friction models [123]. In addition, Costa et al., from the same research group, introduced the idea of using fractal texture profiles for rendering [31]; their virtual stylus model computed a lateral force proportional to the slope of the profile and rendered them on an Impulse Engine 2000; the RMS of the texture profile emerged as the most important factor for the perceived roughness of fractal textures.

Finally, Kuchenbecker improved reality based haptics by introducing the event based haptic method for virtual contacts [81], that is, the haptic device is programmed to replicate the acceleration profile that a user would experience when tapping an object. The open loop models of Kuchenbecker included also the inverse dynamics of the PHANTOM™. The same technique was successfully applied to extend the bandwidth of a teleoperation system [80]. While reality based haptics, as defined by Okamura, can be used over all the workspace of the device, the high quality results achieved by Kuchenbecker are probably confined to small parts of the workspace, because the closed form solution of inverse dynamic model changes with the position of the haptic device.

While the previous works dealt only with the implementation of new algorithms on available hardware, Wall and Harwin added a high fidelity probe to a standard PHANTOM™ haptic device [55]. The probe was a stylus shaped handle with a loudspeaker-type actuator that generated vibrations along the axis of the handle. This probe was meant to overcome the bandwidth limitations of the standard PHANTOM™ haptic device and was shown to improve the detectability of texture's orientation, when compared with the original device [144]. The large bandwidth of this probe (up to 1.5 kHz) allowed a precise representation of frequency based signals, but could not provide enough stiffness to create virtual boundaries as strong as the original PHANTOM™. Moreover, the probe could render vibrations only in 1D, which would be felt along the stylus axis. Nevertheless, Wall and Harwin used the probe to test a new algorithm for texture synthesis based on recorded data. First the geometrical profiles of different textures (mouse mat, sandpaper, and cardboard) were acquired, then the amplitude spectrum of such profiles was computed with the Fourier transform and was used to render virtual textures. To account for inter samples variations of the same materials, the coefficients for the Fourier reconstructions were chosen from a Gaussian distribution. The experiment, carried out with the high frequency probe, showed that, although the subject reported that the virtual surface felt different from the real material, they were able to distinguish between the three materials, confirming the validity of Fourier based stochastic methods.

2.5.3 Stochastic Models

Wall and Harwin were inspired by previous works on stochastic modeling of virtual haptic textures.

Initially, Siira and Pai proposed the idea of modeling textures with stochastic models in the time domain [132]; their system is based on the observation that finished surfaces have normally distributed geometrical features. Since the Gaussian distribution is invariant with respect to resampling, the temporal sequence of surface asperities encountered by the haptic device has the same mean and standard deviation of the spatial distribution of the features; moreover, this result is independent from the user velocity. Pai's algorithm exploited this feature and generated the texture profile as a stochastic sequence in the time domain. The resulting geometric profile is not spatially consistent because different heights can be attributed to the same point on the surface at different times; this algorithm was tested on a Pantograph haptic device and was used to render both a 2D lateral texture as well as a 1D texture over a virtual constraint. In this last case, the lateral force was made proportional to the normal force generated by the virtual constraint.

This assumption of Gaussian distributed features was further investigated by Green and Salisbury in the context of remote sensing of soil properties [52]. A PHANTOM™ based system for acquisition and rendering of virtual textures was used to measure the varying friction coefficient of different sandpaper surfaces; the histogram analysis of the friction showed a Gaussian distribution, whose standard deviation decreased as the grit number grew. A rendering algorithm was also proposed to exploit this statistical property, and it generated a non deterministic, Gaussian, spatially varying friction field, whose mean and standard deviations were computed from the data.

A more complete framework for stochastic textures was proposed by Fritz and Barner [42]; rather than focusing on realism, this method was aimed at generating haptic textures perceptually different from each other. The proposed models were a combination of three different components: deterministic profiles based on either multivariate Gaussians or sums of sine functions, random processes that perturbed such profiles, and banks of bandpass filters that shaped white noise. The perceptual roughness of Gaussian texture was consistent with previous results, where the bigger the variance the rougher the texture. Users reported that Gaussian textures felt like granite. Fritz and Barner also implemented a volumetric version of their texture algorithm, and its exploration felt like moving through gravel in a container.

Afterward, Pai et al. realized a robotic facility aimed at acquiring a complete virtual model of real objects [115]; their system acquires visual, haptic, and auditory measurements and produces physically based representation of the object properties. In particular, their representation of haptic textures was based on a directional varying friction field, which was computed with autoregressive models identified from the data. This method assumes that roughness perception is not dependent on the waveform of the force signal but only on its statistical features.

Finally, granular synthesis of haptic textures explicitly uses stochastic models to generate haptic textures from a discrete set of spatially localized samples (called grains) [32]. Similarly to other stochastic processes: "...the same area of surface will not have exactly the same texture at different points in time although the perceived qualities will be similar." This limitation applies to all methods that use stochastic models to generate time varying textural forces, while it does not concern methods that generate random texture profiles and then use deterministic procedures to compute the forces.

2.5.3.1 Perceived Instability and Artifacts

Since most haptic research has been developed on the PHANTOM™ haptic device, the limitations of the hardware has greatly affected the quality of the results. Apart from the limited mechanical bandwidth of the device, which was already discussed, artifacts can also be introduced by the synthesis method implemented in the virtual environment and by the controller of the device.

The most comprehensive study on perceptual artifacts related to haptic textures with a PHANTOM™ was conducted by Choi and Tan [20, 22, 23]. These studies discussed the perceptual instabilities of two texture rendering algorithms, based on Massie's algorithm [98], and on Ho's [57]. First, the effects of the different algorithms and exploration strategies were studied [15, 17, 19, 20], and three types of instability were described: buzzing, aliveness, and ridge instability. The first term indicates the onset of fast limit cycles due to a non passive control of the device. Ridge instability refers to the feeling of being "sucked" into the textures; this phenomenon is solely related to the texture rendering algorithm, which generates a field independent from the user-exerted force. Lastly aliveness describes a low frequency artifact, different from buzzing, which is caused by large force variations in response of small stylus movements; this last artifact is elusive and could not be pinpointed either in a spectral nor in a temporal analysis of the user interaction with the virtual texture. Ho's algorithm exhibited both buzzing and ridge instability, while Massie's was characterized by buzzing and aliveness. In the same experiments, texture exploration strategies showed a clear effect on the perceived instability. In particular, the action of stroking a texture was less prone to artifacts than free exploration.

In this first round of experiments, Choi and Tan used a variant of Massie's algorithm which generates discontinuous force fields. As a result, low stiffness virtual environment induced non passive behaviors, thus generating the so-called buzzing. Moreover, no mention to Ho's corrective measures was made, when discussing his bump mapping technique.

The discontinuity of the force fields is discussed in [16, 22], where both continuous versions of the algorithms are reported. The continuous version allowed stiffer virtual textures, hence partially resolving the buzzing problem, which is not surprising if examined from the control system point of view. On the other hand, aliveness was proved to be independent from the passivity of the haptic rendering, but a precise physical characterization was not found.

In the last experiments the effect of the update rate of the virtual environment was characterized [18, 21, 23]. Control theory had already found a mathematical relationship between update rate and stiffness of the virtual environment, but this framework is not discussed in Choi and Tan's papers. Moreover, their results are somehow in contrast with control theory, a topic that is not mentioned in their conclusions. Nevertheless, the experiments confirmed the beneficial effects of higher update rates on buzzing; the paper concluded with the observation that higher update rates did not affect the perception of virtual textures, as long as no buzzing was present. According to the authors, the advantages of fast update rates are related only to stability and passivity concerns.

In general, Choi and Tan employed, at most, five subjects in their experiments and sometimes as few as two. The evaluation of their results must take into account the limited data available, as well as the fact that only some of the subjects participated in all the studies. Despite these limitations, this research offers a first important insight on the problem of rendering force feedback haptic textures; however, the strong focus on perceptual qualities and the limited analysis of the engineering aspects of the rendering system does not allow for a quick generalization of these results, because the influence of the haptic device is not sufficiently discussed and the algorithms are not linked to control theory.

Tan's group continued the research on haptic textures with the ministick, Sect. 2.2.2.6 at page 12; their research provided a frequency analysis of perception of virtual haptic textures [25] and on the discriminability of real and virtual textures with such device [77]. In the first of these papers, users were able to discriminate sinusoidal textures from square waves based on the harmonic components of the signal. The second paper discusses the difference in detection thresholds between real and virtual textures: because the thresholds of these two cases is similar they conclude that the ministick is adequate for texture perception studies. While no argument can be made against this claim, few engineering characteristics are provided for the ministick, for example its bandwidth; even the resolution of the device is not clearly indicated (authors claim an astonishing 1 μm resolution in contrast with the 10 μm cited in the design paper). As long as these aspects are not clarified the research carried out with the ministick cannot be readily extended to other devices; in addition, the investigation of the somatosensory system carried out with such device cannot be quickly accepted, due to the unavailability of the device's properties.

2.5.4 Perceived Roughness of Virtual Textures

In a two paper series Lederman and Klatzky assessed the perception of virtual viscous textures. In the first study with a Wingman Haptic mouse (a 2D pantograph style haptic device designed for the video games enthusiast), they found that by varying the resistive force it is possible to elicit the sensation of roughness. The space was partitioned in resistive strips (called ridges) separated by non resistive strips (called grooves), in a pattern mimicking the square gratings used in previous studies. The roughness estimated were completely different from the bare finger and probe case: users probably relied on the total resistance of the surface to rate roughness, but the authors were still able to detect significant effects of the microgeometry. In the second series of experiments, carried out with a PHANTOM™, they confirmed that users rated as rougher surfaces that offered more resistance: the coefficient of viscous friction, in fact, was the single most important effect; but roughness arises in function of the variations in viscosity and not of the mean resistive force. Finally, with the PHANTOM™ the spatial determinants had either not significant or very small effects. Due to the difference in fidelity of the two devices, a direct comparison of the results is difficult.

When exploring the visuo-tactile multimodal perception of virtual roughness, Drewing and colleagues found that the geometric model proposed by Lederman is also valid for virtual textures: when a PHANTOM™ haptic device is used to simulate a point contact on a surface with raised dots, the virtual probe has zero radius hence, according to the model, roughness perception should decrease as inter-element spacing increases, and this is the case [37]. A similar trend was found for virtual sinusoidal gratings both in blind and non-blind subjects [79].

2.6 Conclusions

This literature review shows that textures are a fundamental element for the experience of haptic virtual environments. Although many studies deal with the perceptual properties of real textures, currently available haptic devices are a major obstacle to the investigation of the perception of virtual haptic textures. Because of their specifications, haptic devices can generate stimuli far stronger than the perceptual thresholds identified by research in the psychophysics of touch.

This problem is compounded by the lack of a framework that would express the rendering capabilities of the haptic devices; also, the algorithms used to convert geometric profiles (or measurements) in force fields have not been discussed in the context of passivity/stability, leaving uncertainty on their performance: a notable example are the artifacts reported by Hong Tan and colleagues that, without a clear characterization of the device and the algorithm, cannot be easily interpreted [17].

Most of the passivity/stability results in literature regard the simple example of a linear unidimensional virtual wall; haptic textures are multidimensional, non linear, and possibly non-conservative force fields, a combination that was never investigated by haptic researchers. As a result, the passivity properties of virtual textures are not known.

If these challenges were solved, it would be finally possible to compare the perceptual properties of different algorithms in relationship with the properties of the device used; for example, once passivity is ensured and the correct rendering of the textures is guaranteed, perceptual artifacts could be reliably classified and linked to either the characteristics of the device or to the parameters governing the haptic algorithm, thus simplifying the design process of the virtual environment.

References

1. Abbott, J.J., Okamura, A.M.: Effects of position quantization and sampling rate on virtual wall passivity. IEEE Trans. Robot. **21**(5), 952–964 (2005)
2. Adams, R.J., Hannaford, B.: Stable haptic interaction with virtual environments. IEEE Trans. Robot. Autom. **15**(3), 465–474 (1999)
3. Adelstein, B.D., Ho, P., Kazerooni, H.: Kinematic design of a three degree of freedom parallel hand controller mechanism. In: Proceedings of the ASME Dynamic Systems and Control Division, vol. 58, pp. 539–545 (1996)
4. Basdogan, C., Ho, C.-H., Srinivasan, M.A.: A ray-based haptic rendering technique for displaying shape and texture of 3D objects in virtual environments. In: Proceedings of the Second PHANToM Users Group Workshop, vol. 61, pp. 77–84 (1997)
5. Bensmaïa, S.J., Hollins, M.: Complex tactile waveform discrimination. J. Acoust. Soc. Am. **108**(3), 1236–1245 (2000)
6. Bensmaïa, S., Hollins, M., Yau, J.: Vibrotactile intensity and frequency information in the pacinian system: A psychophysical model. Percept. Psychophys. **67**(5), 828–832 (2005)
7. Birznieks, I., Jenmalm, P., Goodwin, A.W., Johansson, R.S.: Encoding of direction of fingertip forces by human tactile afferents. J. Neurosci. **21**(20), 8222–8237 (2001)
8. Blake, D.T., Johnson, K.O., Hsiao, S.S.: Monkey cutaneous SAI and RA responses to raised and depressed scanned patterns: Effects of width, height, orientation, and a raised surround. J. Neurophysiol. **78**(5), 2503–2517 (1997)
9. Blinn, J.F.: Simulation of wrinkled surfaces. In: SIGGRAPH '78: Proceedings of the 5th Annual Conference on Computer Graphics and Interactive Techniques, pp. 286–292. ACM Press, New York (1978)
10. Bonneton, E., Hayward, V.: Pantograph project, chapter: Implementation of a virtual wall. Technical report, McGill Research Center for Intelligent Machines, McGill University, Montreal, Canada (1994)
11. Campion, G., Wang, Q., Hayward, V.: The Pantograph Mk-II: A haptic instrument. In: Proceedings of the IEEE/RSJ International Conference on Intelligent Robots and Systems, IROS'05, pp. 723–728 (2005)
12. Cascio, C.J., Sathian, K.: Temporal cues contribute to tactile perception of roughness. J. Neurosci. **21**(14), 5289–5296 (2001)
13. Cavusoglu, M.C., Feygin, D., Tendick, F.: A critical study of the mechanical and electrical properties of the PHANToM haptic interface and improvements for high performance control. Presence **11**(6), 555–568 (2002)

14. Chapman, C.E., Tremblay, F., Jiang, W., Belingard, L., Meftah, E.-M.: Central neural mechanisms contributing to the perception of tactile roughness. Behav. Brain Res. **132**(1–2), 225–233 (2002)
15. Choi, S., Tan, H.Z.: An analysis of perceptual instability during haptic texture rendering. In: Proceedings 10th Symposium on Haptic Interfaces for Virtual Environment and Teleoperator Systems. HAPTICS 2002, Orlando, FL, USA, pp. 129–136 (2002)
16. Choi, S., Tan, H.Z.: An experimental study of perceived instability during haptic texture rendering: Effects of collision detection algorithm. In: Proceedings 11th Symposium on Haptic Interfaces for Virtual Environment and Teleoperator Systems. HAPTICS 2003, Los Angeles, CA, USA, pp. 197–204 (2003)
17. Choi, S., Tan, H.Z.: Aliveness: Perceived instability from a passive haptic texture rendering system. In: Proceedings of the IEEE/RSJ International Conference on Intelligent Robots and Systems (IROS), vol. 3, Las Vegas, NV, United States, pp. 2678–2683 (2003)
18. Choi, S., Tan, H.Z.: Effect of update rate on perceived instability of virtual haptic texture. In: Proceedings of the IEEE/RSJ International Conference on Intelligent Robots and Systems (IROS) (IEEE Cat. No.04CH37566), vol. 4, Sendai, Japan, pp. 3577–3582 (2004)
19. Choi, S., Tan, H.Z.: Toward realistic haptic rendering of surface textures. IEEE Comput. Graph. Appl. **24**(2), 40–47 (2004)
20. Choi, S., Tan, H.Z.: Perceived instability of virtual haptic texture. I. Experimental studies. Presence **13**(4), 395–415 (2004)
21. Choi, S., Tan, H.Z.: Discrimination of virtual haptic textures rendered with different update rates. In: Proceedings of the First Joint Eurohaptics Conference and Symposium on Haptic Interfaces for Virtual Environments and Teleoperator Systems WHC'05, Pisa, Italy, pp. 114–119 (2005)
22. Choi, S., Tan, H.Z.: Perceived instability of virtual haptic texture. II. Effect of collision-detection algorithm. Presence **14**(4), 463–481 (2005)
23. Choi, S., Tan, H.Z.: Perceived instability of virtual haptic texture. III. Effect of update rate. Presence **16**(3), 263–278 (2007)
24. Choi, S., Walker, L., Tan, H.Z., Crittenden, S., Reifenberger, R.: Force constancy and its effect on haptic perception of virtual surfaces. ACM Trans. Appl. Percept. **2**(2), 89–105 (2005)
25. Cholewiak, S., Tan, H.Z.: Frequency analysis of the detectability of virtual haptic gratings. In: WHC '07: Proceedings of the Second Joint EuroHaptics Conference and Symposium on Haptic Interfaces for Virtual Environment and Teleoperator Systems, pp. 27–32. IEEE Computer Society, Washington (2007)
26. Colgate, J.E., Brown, J.M.: Factors affecting the Z-width of a haptic display. In: Proceedings of the IEEE International Conference on Robotics and Automation, pp. 3205–3210 (1994)
27. Colgate, J.E., Schenkel, G.: Passivity of a class of sampled-data systems: Application to haptic interfaces. In: Proceedings of the American Control Conference, pp. 3236–3240 (1994)
28. Colgate, J.E., Schenkel, G.G.: Passivity of a class of sampled-data systems: Application to haptic interfaces. J. Robot. Syst. **14**(1), 37–47 (1997)
29. Connor, C., Johnson, K.: Neural coding of tactile texture: Comparison of spatial and temporal mechanisms for roughness perception. J. Neurosci. **12**(9), 3414–3426 (1992)
30. Connor, C., Hsiao, S., Phillips, J., Johnson, K.: Tactile roughness: Neural codes that account for psychophysical magnitude estimates. J. Neurosci. **10**(12), 3823–3836 (1990)
31. Costa, M.A., Cutkosky, M.R.: Roughness perception of haptically displayed fractal surfaces. In: Proceedings ASME IMECE Symposium on Haptic Interfaces for Virtual Environments and Teleoperator Systems, vol. 69-2, pp. 1073–1079 (2000)
32. Crossan, A., Williamson, J., Murray-Smith, R.: Haptic granular synthesis: Targeting, visualisation and texturing. In: Proceedings of the International Symposium on Non-visual & Multimodal Visualization, pp. 527–532. IEEE Press, New York (2004)
33. DiCarlo, J.J., Johnson, K.O., Hsiao, S.S.: Structure of receptive fields in area 3b of primary somatosensory cortex in the alert monkey. J. Neurosci. **18**(7), 2626–2645 (1998)

34. Diolaiti, N., Niemeyer, G., Barbagli, F., Salisbury, J.K.: Stability of haptic rendering: Discretization, quantization, time delay, and coulomb effects. IEEE Trans. Robot. **22**(2), 256–268 (2006)
35. Diolaiti, N., Niemeyer, G., Tanner, N.A.: Wave haptics: Building stiff controllers from the natural motor dynamics. Int. J. Robot. Res. **26**(1), 5–21 (2007)
36. Dodson, M.J., Goodwin, A.W., Browning, A.S., Gehring, H.M.: Peripheral neural mechanisms determining the orientation of cylinders grasped by the digits. J. Neurosci. **18**(1), 521–530 (1998)
37. Drewing, K., Ernst, M.O., Lederman, S.J., Klatzky, R.: Roughness and spatial density judgments on visual and haptic textures using virtual reality. In: Proceedings of EuroHaptics (2004)
38. Ekman, G., Hosman, J., Lindström, B.: Roughness, smoothness, and preference: A study of quantitative relations in individual subjects. J. Exp. Psychol. **70**(1), 18–26 (1965)
39. Force Dimension: Delta. http://www.forcedimension.com/delta3-overview
40. Force Dimension: Omega. http://www.forcedimension.com/omega3-overview
41. Frisoli, A., Rocchi, F., Marcheschi, S., Dettori, A., Salsedo, F., Bergamasco, M.: A new force-feedback arm exoskeleton for haptic interaction in virtual environments. In: Proceedings of the First Joint Eurohaptics Conference and Symposium on Haptic Interfaces for Virtual Environments and Teleoperator Systems WHC'05, pp. 195–201 (2005)
42. Fritz, J.P., Barner, K.E.: Stochastic models for haptic textures. In: Stein, M.R. (ed.) Telemanipulator and Telepresence Technologies III. Proc. SPIE, vol. 2901, pp. 34–44 (1996)
43. Gescheider, G.A., Bolanowski, S.J., Greenfield, T.C., Brunette, K.E.: Perception of the tactile texture of raised-dot patterns: A multidimensional analysis. Somatosens. Motor Res. **22**(3), 127–140 (2005)
44. Gibson, J.J.: Observations on active touch. Psychol. Rev. **69**, 477–491 (1962)
45. Gil, J.J., Sanchez, E., Hulin, T., Preusche, C., Hirzinger, G.: Stability boundary for haptic rendering: Influence of damping and delay. In: Proceedings of the IEEE International Conference on Robotics and Automation, pp. 124–129 (2007)
46. Gillespie, B., Cutkosky, M.: Stable user-specific rendering of the virtual wall. In: Proceedings of the ASME Dynamic Systems and Control Division, vol. DSC-Vol. 58, pp. 397–406 (1996)
47. Goodwin, A., John, K., Sathian, K., Darian-Smith, I.: Spatial and temporal factors determining afferent fiber responses to a grating moving sinusoidally over the monkey's fingerpad. J. Neurosci. **9**(4), 1280–1293 (1989)
48. Goodwin, A., Browning, A., Wheat, H.: Representation of curved surfaces in responses of mechanoreceptive afferent fibers innervating the monkey's fingerpad. J. Neurosci. **15**(1), 798–810 (1995)
49. Goodwin, A.W., Wheat, H.E.: Effects of nonuniform fiber sensitivity, innervation geometry, and noise on information relayed by a population of slowly adapting type I primary afferents from the fingerpad. J. Neurosci. **19**(18), 8057–8070 (1999)
50. Goodwin, A.W., Macefield, V.G., Bisley, J.W.: Encoding object curvature by tactile afferents from human fingers. J. Neurophysiol. **78**, 2881–2888 (1997)
51. Gordon, I.E., Cooper, C.: Improving one's touch. Nature **256**, 203–204 (1975)
52. Green, D.F., Salisbury, J.K.: Texture sensing and simulation using the PHANToM: Towards remote sensing of soil properties. In: Proceedings of the Second Phantom Users Group Workshop (1997)
53. Hannaford, B., Ryu, J.H.: Time-domain passivity control of haptic interfaces. IEEE Trans. Robot. Autom. **18**(1), 1–10 (2002)
54. Hardwick, A., Furner, S., Rush, J.: Tactile display of virtual reality from the world wide web—a potential access method for blind people. Displays **18**, 153–161 (1998)
55. Harwin, W.S., Wall, S.A.: Mechatronic design of a high frequency probe for haptic interaction. In: Proceedings 6th International Conference on Mechatronics and Machine Vision in Practice, pp. 111–118 (1999)
56. Hayward, V., Yi, D.: Change of height: An approach to the haptic display of shape and texture without surface normal. In: Siciliano, B., Dario, P. (eds.) Experimental Robotics VIII. Springer Tracts in Advanced Robotics, pp. 570–579. Springer, Heidelberg (2003)

57. Ho, C.-H., Basdogan, C., Srinivasan, M.A.: Efficient point-based rendering techniques for haptic display of virtual objects. Presence **8**(5), 477–491 (1999)
58. Ho, P.P., Adelstein, B.D., Kazerooni, H.: Judging 2D versus 3D square-wave virtual gratings. In: Proceedings of the 12th International Symposium on Haptic Interfaces for Virtual Environment and Teleoperator Systems, pp. 176–183 (2004)
59. Hollins, M., Bensmaia, S.J.: The coding of roughness. Can. J. Exp. Psychol. **61**(3), 184–195 (2007)
60. Hollins, M., Risner, S.R.: Evidence for the duplex theory of tactile texture perception. Percept. Psychophys. **62**(4), 695–705 (2000)
61. Hollins, M., Faldowski, R., Rao, S., Young, F.: Perceptual dimensions of tactile surface texture: A multidimensional scaling analysis. Percept. Psychophys. **54**, 697–705 (1993)
62. Hollins, M., Fox, A., Bishop, C.: Imposed vibration influences perceived tactile smoothness. Perception **29**, 1455–1465 (2000)
63. Hollins, M., Bensmaïa, S.J., Karlof, K., Young, F.: Individual differences in perceptual space for tactile textures: Evidence from multidimensional scaling. Percept. Psychophys. **62**(8), 1534–1544 (2000)
64. Hollins, M., Bensmaïa, S.J., Washburn, S.: Vibrotactile adaptation impairs discrimination of fine, but not coarse, textures. Somatosens. Motor Res. **18**(10), 253–262 (2001)
65. Hollins, M., Lorenz, F., Harper, D.: Somatosensory coding of roughness: The effect of texture adaptation in direct and indirect touch. J. Neurosci. **26**(20), 5582–5588 (2006)
66. Hulin, T., Preusche, C., Hirzinger, G.: Stability boundary for haptic rendering: Influence of physical damping. In: Intelligent Robots and Systems, 2006 IEEE/RSJ International Conference on, pp. 1570–1575 (2006)
67. Immersion Corporation: Impulse Engine 2000. http://www.immersion.com/
68. Immersion Corporation: Impulse Stick. http://www.immersion.com/industrial/joystick/impulse_stick.php
69. Immersion Corporation: Rotary Haptic Knob. http://www.immersion.com/products/touchsense-force-feedback/6000-series/rotary.html
70. Johnson, K.O.: The roles and functions of cutaneous mechanoreceptors. Curr. Opin. Neurobiol. **11**(4), 455–461 (2001)
71. Johnson, K.O., Hsiao, S.S.: Neural mechanisms of tactual form and texture perception. Annu. Rev. Neurosci. **15**(1), 227–250 (1992)
72. Johnson, K.O., Yoshioka, T., Vega-Bermudez, F.: Tactile functions of mechanoreceptive afferents innervating the hand. J. Clin. Neurophysiol. **17**, 539–558 (2000)
73. Klatzky, R.L., Lederman, S.J.: Tactile roughness perception with a rigid link interposed between skin and surface. Percept. Psychophys. **61**(4), 591–607 (1999)
74. Klatzky, R.L., Lederman, S.J.: The perceived roughness of resistive virtual textures: I. Rendering by a force-feedback mouse. ACM Trans. Appl. Percept. **3**(1), 1–14 (2006)
75. Klatzky, R.L., Lederman, S.J., Hamilton, C., Grindley, M., Swendsen, R.H.: Feeling textures through a probe: Effects of probe and surface geometry and exploratory factors. Percept. Psychophys. **65**, 613–631 (2003)
76. Knibestol, M., Vallbo, A.B.: Intensity of sensation related to activity of slowly adapting mechanoreceptive units in the human hand. J. Physiol. **300**(1), 251–267 (1980)
77. Kocsis, M., Tan, H.Z., Adelstein, B.D.: Discriminability of real and virtual surfaces with triangular gratings. In: WHC '07: Proceedings of the Second Joint EuroHaptics Conference and Symposium on Haptic Interfaces for Virtual Environment and Teleoperator Systems, pp. 348–353. IEEE Computer Society, Washington (2007)
78. Kontarinis, D.A., Howe, R.D.: Tactile display of vibratory information in teleoperation and virtual environments. Presence **4**(4), 387–402 (1995)
79. Kornbrot, D., Penn, P., Petrie, H., Furner, S., Hardwick, A.: Roughness perception in haptic virtual reality for sighted and blind people. Percept. Psychophys. **69**(4), 502–512 (2007)
80. Kuchenbecker, K.J., Niemeyer, G.: Improving telerobotic touch via high-frequency acceleration matching. In: Proceedings of the IEEE Int. Conf. on Robotics and Automation, 2006
81. Kuchenbecker, K.J., Fiener, J., Niemeyer, G.: Improving contact realism through event-based haptic feedback. IEEE Trans. Vis. Comput. Graph. **12**(2), 219–230 (2006)

82. LaMotte, R.H., Whitehouse, J.: Tactile detection of a dot on a smooth surface: Peripheral neural events. J. Neurophysiol. **56**, 1109–1128 (1986)
83. Lawrence, M.A., Kitada, R., Klatzky, R.L., Lederman, S.J.: Haptic roughness perception of linear gratings via bare finger or rigid probe. Perception **36**(4), 547–557 (2007)
84. Lederman, S., Klatzky, R., Hamilton, C., Grindley, M.: Perceiving surface roughness through a probe: Effects of applied force and probe diameter. In: Proceedings of the ASME DSCD-IMECE (2000)
85. Lederman, S.J.: Tactile roughness of grooved surfaces: The touching process and effects of macro- and microsurface structure. Percept. Psychophys. **16**(2), 385–395 (1974)
86. Lederman, S.J.: "improving one's touch" ... and more. Percept. Psychophys. **24**(2), 154–160 (1978)
87. Lederman, S.J.: The perception of surface roughness by active and passive touch. Bull. Psychon. Soc. **18**(5), 253–255 (1981)
88. Lederman, S.J.: Tactual roughness perception: Spatial and temporal determinants. Can. J. Psychol. **37**(4), 498–511 (1983)
89. Lederman, S.J., Taylor, M.M.: Fingertip force, surface geometry, and the perception of roughness by active touch. Percept. Psychophys. **12**, 401–408 (1972)
90. Lederman, S.J., Loomis, J.M., Williams, D.A.: The role of vibration in tactual perception of roughness. Percept. Psychophys. **32**(2), 109–116 (1982)
91. Lederman, S.J., Klatzky, R.L., Hamilton, C.L., Ramsay, G.I.: Perceiving roughness via a rigid probe: Psychophysical effects of exploration speed and mode of touch. Haptics-E: Electron. J. Haptics Res. **1** (1999), online
92. Lederman, S.J., Klatzky, R.L., Tong, C., Hamilton, C.: The perceived roughness of resistive virtual textures: II. Effects of varying viscosity with a force-feedback device. ACM Trans. Appl. Percept. **3**(1), 15–30 (2006)
93. Levesque, V., Hayward, V.: Tactile graphics rendering using three laterotactile drawing primitives. In: Proceedings of the Symposium on Haptic Interfaces for Virtual Environment and Teleoperator Systems, pp. 429–436 (2008)
94. Logitech: G27 Racing Wheel. http://www.logitech.com/en-ca/gaming/wheels
95. Louw, S., Kappers, A.M.L., Koenderink, J.J.: Haptic detection thresholds of Gaussian profiles over the whole range of spatial scales. Exp. Brain Res. **132**, 369–374 (2000)
96. MacLean, K.E., Snibbe, S.S.: An architecture for haptic control of media. In: Proceedings of the Symposium on Haptic Interfaces for Virtual Environment and Teleoperator Systems, pp. 219–228 (1999)
97. Mahvash, M., Hayward, V.: High fidelity passive force reflecting virtual environments. IEEE Trans. Robot. **21**(1), 38–46 (2005)
98. Massie, T.H.: Initial haptic explorations with the PHANToM virtual touch through point interaction. Master's thesis, Massachusetts Institute of Technology (1996)
99. Massie, T.H., Salisbury, J.K.: The PHANToM haptic interface: A device for probing virtual objects. In: Proceedings ASME IMECE Symposium on Haptic Interfaces for Virtual Environments and Teleoperator Systems, vol. DSC-Vol. 55-1, pp. 295–301 (1994)
100. Meenes, M., Zigler, M.J.: An experimental study of the perceptions roughness and smoothness. Am. J. Psychol. **34**(4), 542–549 (1923)
101. Meftah, E.M., Belingard, L., Chapman, C.E.: Relative effects of the spatial and temporal characteristics of scanned surfaces on human perception of tactile roughness using passive touch. Exp. Brain Res. **132**(3), 351–361 (2000)
102. Miller, B.E., Colgate, J.E., Freeman, R.A.: Guaranteed stability of haptic systems with nonlinear virtual environments. IEEE Trans. Robot. Autom. **16**(6), 712–719 (2000)
103. Minsky, M.: Computational haptics: The sandpaper system for synthesizing texture for a force-feedback display. PhD thesis, Massachusetts Institute of Technology (1995)
104. Minsky, M., Lederman, S.J.: Simulated haptic textures: Roughness. In: Proceedings of the ASME IMECE Symposium on Haptic Interfaces for Virtual Environments and Teleoperator Systems, DSC-Vol. 58, pp. 421–426 (1996)

105. Minsky, M., Ming, O., Steele, O., Brooks, Jr., F.P., Behensky, M.: Feeling and seeing: Issues in force display. In: Proceedings of Conference on Computer Graphics and Interactive Techniques, SIGGRAPH'90, vol. 24(2), pp. 235–241 (1990)
106. Morley, J.W., Goodwin, A.W., Darian-Smith, I.: Tactile discrimination of gratings. Exp. Brain Res. **49**(2), 291–299 (1983)
107. Nefs, H.T., Kappers, A.M.L., Koenderink, J.J.: Amplitude and spatial-period discrimination in sinusoidal gratings by dynamic touch. Perception **30**, 1263–1274 (2001)
108. Niemeyer, G., Slotine, J.-J.: Stable adaptive teleoperation. IEEE J. Ocean. Eng. **16**(1), 152–162 (1991)
109. Niemeyer, G., Slotine, J.-J.E.: Telemanipulation with time delays. Int. J. Robot. Res. **23**(9), 873–890 (2004)
110. Okamura, A., Dennerlein, J.T., Howe, R.D.: Vibration feedback models for virtual environments. In: Proceedings of IEEE International Conference on Robotics and Automation, vol. 3, pp. 2485–2490 (1998)
111. Okamura, A., Hage, M., Dennerlein, J., Cutkosky, M.: Improving reality-based models for vibration feedback. In: Proceedings of the ASME Dynamic Systems and Control Division, vol. 69, pp. 1117–1124 (2000)
112. Okamura, A.M., Cutkosky, M.R., Dennerlein, J.T.: Reality-based models for vibration feedback in virtual environments. IEEE/ASME Trans. Mechatron. **6**(3), 245–252 (2001)
113. Okamura, A.M., Costa, M.A., Turner, M.L., Richard, C., Cutkosky, M.R.: Haptic surface exploration. In: The Sixth International Symposium on Experimental Robotics VI, pp. 423–432. Springer, London (2000)
114. Otaduy, M.A., Lin, M.C.: A perceptually-inspired force model for haptic texture rendering. In: Proceedings of the 1st Symposium on Applied Perception in Graphics and Visualization, pp. 123–126. ACM Press, New York (2004)
115. Pai, D.K., van den Doel, K., James, D.L., Lang, J., Lloyd, J.E., Richmond, J.L., Yau, S.H.: Scanning physical interaction behavior of 3D objects. In: SIGGRAPH '01: Proceedings of the 28th Annual Conference on Computer Graphics and Interactive Techniques, pp. 87–96. ACM Press, New York (2001)
116. Paré, M., Elde, R., Mazurkiewicz, J.E., Smith, A.M., Rice, F.L.: The Meissner corpuscle revised: A multiafferented mechanoreceptor with nociceptor immunochemical properties. J. Neurosci. **21**(18), 7236–7246 (2001)
117. Paré, M., Smith, A.M., Rice, F.L.: Distribution and terminal arborizations of cutaneous mechanoreceptors in the glabrous finger pads of the monkey. J. Comp. Neurol. **445**, 347–359 (2002)
118. Paré, M., Behets, C., Cornu, O.: Paucity of presumed Ruffini Corpuscles in the index fingerpad of humans. J. Comp. Neurol. **356**, 260–266 (2003)
119. Phillips, J.R., Johnson, K.O.: Tactile spatial resolution: II. Neural representation of bars, edges, and gratings in monkey primary afferents. J. Neurophysiol. **46**, 1192–1203 (1981)
120. Picard, D., Dacremont, C., Valentin, D., Giboreau, A.: Perceptual dimensions of tactile textures. Acta Psychol. **114**, 165–184 (2003)
121. Quanser: Haptic Devices. http://www.quanser.com/
122. Ramstein, C., Hayward, V.: The pantograph: A large workspace haptic device for a multimodal human-computer interaction. In: Proceedings of the SIGCHI Conference on Human Factors in Computing Systems, CHI'04, ACM/SIGCHI Companion-4/94, pp. 57–58 (1994)
123. Richard, C., Cutkosky, M., MacLean, K.E.: Friction identification for haptic display. In: Proceedings of the 8th Ann. Symp. on Haptic Interfaces for Virtual Environment and Teleoperator Systems. ASME/IMECE, London (1999)
124. Richard, C., Okamura, A.M., Cutkosky, M.R.: Getting a feel for dynamics: Using haptic interface kits for teaching dynamics and controls. In: American Society of Mechanical Engineers, Dynamic Systems and Control Division, vol. 61, pp. 153–157 (1997)
125. Rosen, M.J., Adelstein, B.D.: Design of a two degree-of-freedom manipulandum for tremor research. In: Proceedings of the IEEE Frontiers of Engineering and Computing in Health Care, pp. 47–51 (1984)

126. Ryu, J.-H., Kwon, D.-S., Hannaford, B.: Control of a flexible manipulator with noncollocated feedback: Time-domain passivity approach. IEEE Trans. Robot. **20**(4), 776–780 (2004)
127. Ryu, J.-H., SangKim, Y., Hannaford, B.: Sampled- and continuous-time passivity and stability of virtual environments. IEEE Trans. Robot. **20**, 772–776 (2004)
128. Ryu, J.H., Preusche, C., Hannaford, B., Hirzinger, G.: Time domain passivity control with reference energy following. IEEE Trans. Control Syst. Technol. **13**(5), 737–742 (2005)
129. Salcudean, S.E., Vlaar, T.D.: On the emulation of stiff walls and static friction with a magnetically levitated inputoutput device. J. Dyn. Syst. **119**(127–132), 127–132 (1997)
130. Sathian, K., Goodwin, A.W., John, K.T., Darian-Smith, I.: Perceived roughness of a grating: Correlation with responses of mechanoreceptive afferents innervating the monkey's fingerpad. J. Neurosci. **9**(4), 1273–1279 (1989)
131. Sensable: Phantom. http://www.sensable.com/products-haptic-devices.htm
132. Siira, J., Pai, D.K.: Haptic textures—a stochastic approach. In: Proceedings of IEEE International Conference on Robotics and Automation, pp. 557–562 (1996)
133. Sirouspour, M.R., DiMaio, S.P., Salcudean, S.E., Abolmaesumi, P., Jones, C.: Haptic interface control-design issues and experiments with a planar device. In: Proceedings of IEEE International Conference on Robotics and Automation, vol. 1, pp. 789–794 (2000)
134. Smith, A.M., Chapman, C.E., Deslandes, M., Langlais, J.S., Thibodeau, M.P.: Role of friction and tangential force variation in the subjective scaling of tactile roughness. Exp. Brain Res. **144**(2), 211–223 (2002)
135. Srinivasan, M.A., Whitehouse, J.M., LaMotte, R.H.: Tactile detection of slip: Surface microgeometry and peripheral neural codes. J. Neurophysiol. **63**(6), 1323–1332 (1990)
136. Steger, R., Lin, K., Adelstein, B.D., Kazerooni, H.: Design of a passively balanced spatial linkage haptic interface. ASME J. Mech. Des. **126**, 984–991 (2004)
137. Stevens, J.C., Harris, J.R.: The scaling of subjective roughness and smoothness. J. Exp. Psychol. **64**, 498–494 (1962)
138. Talbot, W.H., Darian-Smith, I., Kornhuber, H.H., Mountcastle, V.B.: The sense of flutter-vibration: Comparison of the human capacity with response patterns of mechanoreceptive afferents from the monkey hand. J. Neurophysiol. **31**(2), 301–334 (1968)
139. Tan, H.Z., Adelstein, B.D., Traylor, R., Kocsis, M., Hirleman, E.D.: Discrimination of real and virtual high-definition textured surfaces. In: HAPTICS '06: Proceedings of the Symposium on Haptic Interfaces for Virtual Environment and Teleoperator Systems, p. 1. IEEE Computer Society, Washington (2006)
140. Taylor, M.M., Lederman, S.J.: Tactile roughness of grooved surfaces: A model and the effect of friction. Percept. Psychophys. **17**(1), 23–26 (1975)
141. Tiest, W.M.B., Kappers, A.M.L.: Analysis of haptic perception of materials by multidimensional scaling and physical measurements of roughness and compressibility. Acta Psychol. **121**(1), 1–20 (2006)
142. Unger, B., Hollis, R., Klatzky, R.: The geometric model for perceived roughness applies to virtual textures. In: Proceedings of the Symposium on Haptic Interfaces for Virtual Environment and Teleoperator Systems (2008)
143. Vega-Bermudez, F., Johnson, K.O.: SA1 and RA receptive fields, response variability, and population responses mapped with a probe array. J. Neurophysiol. **81**(6), 2701–2710 (1999)
144. Wall, S.A., Harwin, W.S.: Effects of physical bandwidth on perception of virtual gratings. In: Proceedings of the Symposium on Haptic Interfaces for Virtual Environments and Teleoperators, ASME Dynamic Systems and Control Division, pp. 1033–1039 (2000)
145. Weisenberger, J.M., Krier, M.J., Rinker, M.A., Kreidler, S.M.: The role of the end-effector in the perception of virtual surfaces presented via force-feedback haptic devices. In: Proceedings of the ASME Dynamic Systems and Control Division (1999)
146. Weisenberger, J.M., Kreier, M.J., Rinker, M.A.: Judging the orientation of sinusoidal and square-wave virtual gratings presented via 2-DOF and 3-DOF haptic interfaces. Haptics-E **1**(4) (2000), online
147. West, A.M., Cutkosky, M.R.: Detection of real and virtual fine surface features with a haptic interface and stylus. In: Proceedings of the ASME Intl. Mech. Eng. Congress: Dynamic

Systems and Control Division (Haptic Interfaces for Virtual Environments and Teleoperator Systems), vol. DSC-Vol. 61, pp. 159–166 (1997)

148. Wheat, H.E., Goodwin, A.W.: Tactile discrimination of gaps by slowly adapting afferents: Effects of population parameters and anisotropy in the fingerpad. J. Neurophysiol. **84**(3), 1430–1444 (2000)
149. Witney, A.G., Wing, A., Thonnard, J.-L., Smith, A.M.: The cutaneous contribution to adaptive precision grip. Trends Neurosci. **27**, 637–643 (2004)
150. Yoshioka, T., Bensmaïa, S.J., Craig, J.C., Hsiao, S.S.: Texture perception through direct and indirect touch: An analysis of perceptual space for tactile textures in two modes of exploration. Somatosens. Motor Res. **24**, 53–70 (2007)
151. Zigler, M.J.: Review of Katz "Der Augbau der Tastwelt"! Psychol. Bull. **23**, 326–336 (1926)

Chapter 3
The Pantograph Mk-II: A Haptic Instrument

Abstract We describe the redesign and the performance evaluation of a high-performance haptic device system called the Pantograph. The device is based on a two degree-of-freedom parallel mechanism which was designed for optimized dynamic performance, but which also is kinematically well conditioned. The results show that the system is capable of producing accurate tactile signals in the **DC**–400 Hz range and can resolve displacements of the order of 10 μm. Future improvements are discussed.

3.1 Preface to Chap. 3

This chapter presents a radical improvement of the Pantrograph force feedback haptic device, which is used in the following chapters of the book to implement texture rendering algorithms and to test their properties. The most important characteristics of this improved interface are high spatial and temporal resolution, and large acceleration bandwidth. At the time of publication, the Pantograph was the first haptic device capable of guaranteeing the precise rendering of haptic textures up to 400 Hz. To achieve this goal, low-pass filters have been applied to the motor commands, obtaining an effect similar to the equalization of an audio signal; the resulting texture signal is extremely smooth as a result of the oversampling and filtering approach, which, for the first time, was introduced to the haptic community in this paper.

The combination of the filtering approach and the hardware design presented in this chapter are the foundation of the experiments presented in the following chapters; moreover, the verification of rendered acceleration is necessary for the validation of the psychophysic studies presented at the end of the book.

3.1.1 Contributions of Authors

Campion Gianni designed and implemented the control of the haptic device: kinematics, filters, rendering algorithm, and initial calibration procedure. He also pro-

Reprinted from Gianni Campion, Qi Wang, and Vincent Hayward, "The Pantograph Mk-II: A Haptic Instrument." *Proc. IROS 2005, IEEE/RSJ Int. Conf. Intelligent Robots and Systems*, pp. 723–728.

G. Campion, *The Synthesis of Three Dimensional Haptic Textures: Geometry, Control, and Psychophysics*, Springer Series on Touch and Haptic Systems,
DOI 10.1007/978-0-85729-576-7_3, © IEEE 2005

vided the figures for the resolution and isotropy of the device. Qi Wang designed and assembled both the mechanical system and the amplifier system, and assisted in the development of the control. Both authors contributed to the writing of the manuscript. Prof. Hayward supervised the work and edited the manuscript and the figures.

3.2 Introduction

The scientific study of touch, the design of computational methods to synthesize tactile signals, studies in the control of haptic interfaces, the development of force reflecting virtual environments, and other activities, all require the availability of devices that can produce reliable haptic interaction signals. In some cases, it is needed to produce well controlled stimuli. In other cases, it is important to have the knowledge of the structural dynamics of a device, but in all cases, these activities entail having devices which have well characterized signal transfer properties.

Following SensAble's Phantom® and Immersion's Impulse Engine®, several new commercially available general-purpose haptic devices have been recently introduced: MPB's Freedom-6S®, Force Dimension's Omega®, Haption's Virtuose®, Immersion Canada's PenCat/Pro®; plus other application-specific devices. In addition, low-complexity, interesting, high-performance devices have also become available, either from research institutions or from commercial sources [9, 10, 20, 21]. We felt, nonetheless, that a general purpose laboratory system having high performance features, would be a valuable tool.

With this in mind, we set out to revisit the design of the 'Pantograph' haptic device, first publicly demonstrated at the 1994 ACM SIGCHI conference in Boston, MA [22]. Our first goal was the creation of an open architecture system which could be easily replicated from blueprints and from a list of off-the-shelf components. The second goal was to obtain a system which would have superior and known performance characteristics so that it could be used as a scientific instrument. Our intention is to make the system available in open-source, hardware and software, in the near future.

An important aspect of the Pantograph, a planar parallel mechanism (Fig. 3.1d), is the nature of its interface: a non-slip plate on which the fingerpad rests. Judiciously programmed tangential interaction forces f_T at the interface (Fig. 3.1e) have the effect of causing fingertip deformations and tactile sensations that resemble exploring real surfaces.

3.3 Components

3.3.1 Mechanical Structure

The mechanical design has not changed from the original device. The dimensions, as well as the shape of the links, were determined from *dynamic performance* con-

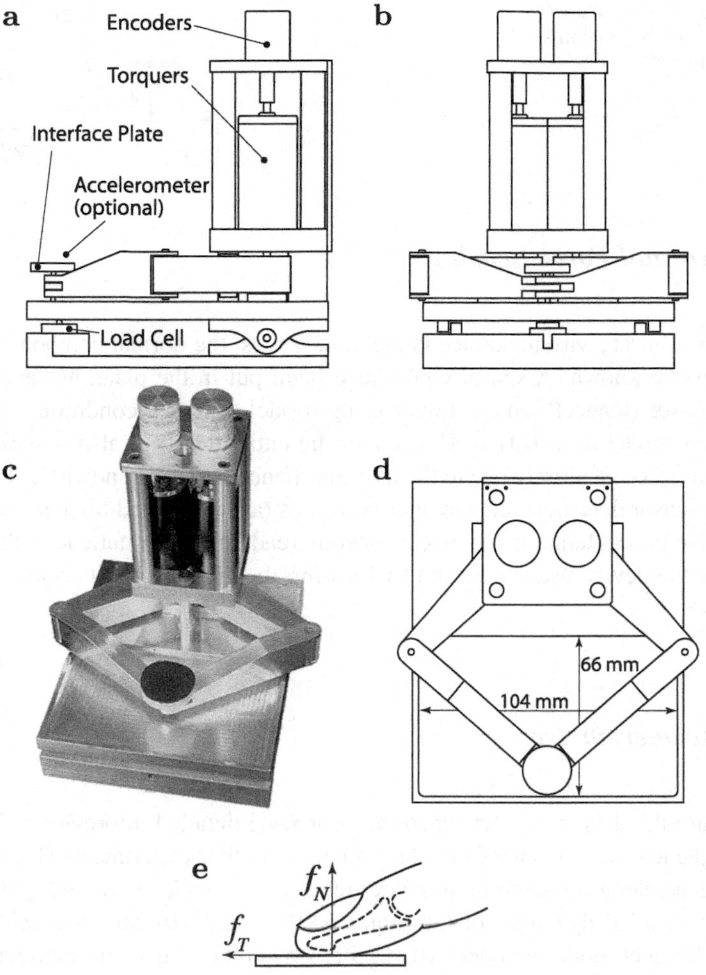

Fig. 3.1 Pantograph Mk II electromechanical hardware. (**a**) Side view showing the main electromechanical components. (**b**) Front view. (**c**) Photograph. (**d**) Top view of the five-bar mechanism and plate constrained to 2-DOF. (**e**) The interaction force has two components: f_N is measured by the load cell and f_T results from coupling the finger tip to the actuators via the linkages

siderations [13], rather than from kinetostatic considerations [24]. Statically, the structure must resist bending when loaded vertically. The proximal links (Fig. 3.2a) have a pocketed box design which gives them the structure of a wishbone horizontally where they are dynamically loaded and otherwise of a hollow beam for torsional static strength. The distal links (Fig. 3.2b) have an axial dynamic load and behave by cantilevers under the vertical static load, therefore they have a tapered shape to reduce weight.

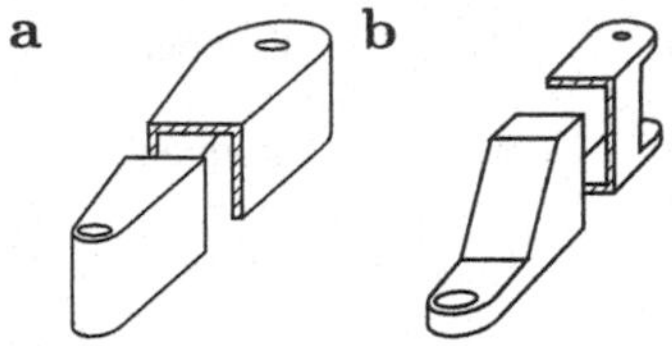

Fig. 3.2 Internal structure of the beams. (**a**) Proximal link. (**b**) Distal link

3.3.2 Normal Force Sensing

To render arbitrary virtual surface interaction forces, the normal component of the force must be known. A sensor could have been put in the plate, however, locating the sensor (loadcell Omega Engineering model LCKD-5; conditioner Newport Electronics model INFS-0010-DC-7) under the entire device is also possible, since the normal force is entirely due to the user and hence has low bandwidth. This way, the force sensor does not 'see' any inertial forces (a tip mounted force sensor could be sensitive to acceleration and give erroneous readings). The static load due to the weight of the device was eliminated by locating the hinge under its center of mass (Fig. 3.1a).

3.3.3 Accelerometer

To measure the device transfer function, to provide detailed information about the high-frequency movements of the plate for use in other experiments (for example involving acceleration feedback to render textures or shock sensations, or to investigate the coupled dynamics of the finger pad), a dual-axis MEMS accelerometer (Analog Device; model number: ADXL250) was embedded in an interchangeable plate (Fig. 3.1a).

3.3.4 Motors

Two conventional coreless DC motors (Maxon RE-25 graphite brushes) are used as torquers. Although this solution is clearly suboptimal, it was used for simplicity and will be further discussed in the Sect. 3.6. We experimented with both graphite and metal brushes. The friction due to metal brushes is lower, but the electrical coupling they provide with the windings at low speeds is not as good as with graphite brushes. It was observed that the electrical resistance varied so greatly and so rapidly from one commutator blade to the next that current feedback was ineffective to compensate for this variation, resulting in noticeable transient drops in the torque.

3.3.5 Position Sensors

The servo quality potentiometers used in the original Pantograph could only provide 10 bits of resolution over the workspace if their signal was unprocessed. These were replaced by optical rotary incremental encoders. Two models were evaluated that had the required resolution and form factor. Models from MicroE Systems Inc. (model M1520S-40-R1910-HA 100,000 CPR) and Gurley Precision Instruments Inc. (model R119S01024Q5L16B188P04MN 65,536 CPR) both gave good results. The Gurley sensors are less expensive and easier to commission while the MicroE sensors require alignment and protective custom housing.

3.3.6 Electronics

An integrated 4-channel "hardware-in-the-loop" PCI card from Quanser Inc. (model Q4) with 24-bit encoder counters, unbuffered, low delay analog-to-digital/digital-to-analog channels proved to be a convenient and cost effective solution (read encoders, read acceleration and force signal, write actuator currents) that could support two devices. The current amplifier design is crucial given the observed variation of the motor winding resistance due to commutation. Low gain current amplifiers built around the NS power chip LM12CL proved to be only partially effective. Better performance should be provided in the future by Quanser's LCAM amplifiers.

3.4 Kinematics

The kinematic structure is a five-bar planar linkage represented in Fig 3.3. The end-plate is located at point $\mathbf{P}_3$ and moves in a plane with two degree-of-freedom with respect to the ground link, where the actuators and sensors are located at $\mathbf{P}_1$ and $\mathbf{P}_5$. The configuration of the device is determined by the position of the two angles θ_1 and θ_5 and the force at the tool tip $\mathbf{P}_3$ is due to torques applied at joints 1 and 5.

The nominal values of the link lengths a_i are in mm:

$$a_{\text{nom}} = [63 \quad 75 \quad 75 \quad 63 \quad 25]^{\top}.$$

3.4.1 Direct Kinematics

The direct kinematics problem consists of finding the position of point $\mathbf{P}_3$ from the two sensed joint angles θ_1 and θ_5. The base frame is set so that its z axis passes through $\mathbf{P}_1$. It was in the past solved using various approaches, the latest provided in [6]. These approaches all share the observation that $\mathbf{P}_3$ is intersection of two

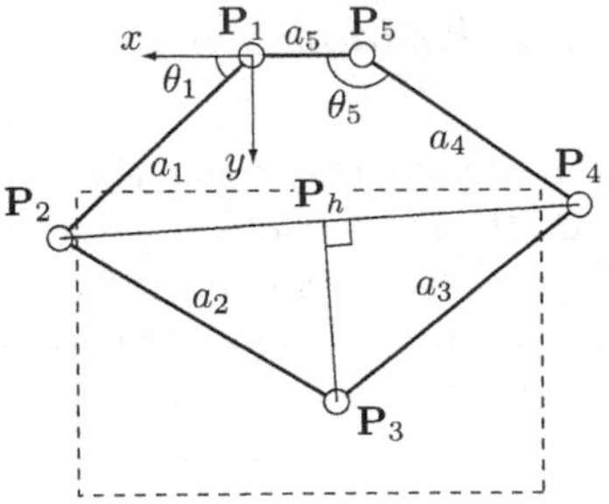

Fig. 3.3 Model of the kinematics used to compute the direct problem

circles, the centers and the radii of which are known. The circles of radii a_2 and a_3 are centered at:

$$\mathbf{P}_2(x_2, y_2) = [a_1\cos(\theta_1),\ a_1\sin(\theta_1)]^\top, \quad \text{and} \tag{3.1}$$

$$\mathbf{P}_4(x_4, y_4) = [a_4\cos(\theta_5) - a_5,\ a_4\sin(\theta_5)]^\top \tag{3.2}$$

and intersect at two points corresponding to two configurations. The device, however, always operates in the configuration that has the largest y. We used a geometric approach to find them.

Let $\mathbf{P}_3 = (x_3, y_3)$ and $\mathbf{P}_h = (x_h, y_h)$ be the intersection between the segment $\mathbf{P}_2\mathbf{P}_4$ and the height of the triangle $\mathbf{P}_2\mathbf{P}_3\mathbf{P}_4$. We find

$$\|\mathbf{P}_2 - \mathbf{P}_h\| = \frac{(a_2^2 - a_3^2 + \|\mathbf{P}_4 - \mathbf{P}_2\|^2)}{(2\|\mathbf{P}_4 - \mathbf{P}_2\|)}, \tag{3.3}$$

$$\mathbf{P}_h = \mathbf{P}_2 + \frac{\|\mathbf{P}_2 - \mathbf{P}_h\|}{\|\mathbf{P}_2 - \mathbf{P}_4\|}(\mathbf{P}_4 - \mathbf{P}_2), \tag{3.4}$$

$$\|\mathbf{P}_3 - \mathbf{P}_h\| = \sqrt{a_2^2 - \|\mathbf{P}_2 - \mathbf{P}_h\|^2}. \tag{3.5}$$

The end effector position $\mathbf{P}_3(x_3, y_3)$ is then given by

$$x_3 = x_h \pm \frac{\|\mathbf{P}_3 - \mathbf{P}_h\|}{\|\mathbf{P}_2 - \mathbf{P}_4\|}(y_4 - y_2), \tag{3.6}$$

$$y_3 = y_h \mp \frac{\|\mathbf{P}_3 - \mathbf{P}_h\|}{\|\mathbf{P}_2 - \mathbf{P}_4\|}(x_4 - x_2). \tag{3.7}$$

The useful solution has a positive sign in Eq. (3.6) and negative sign in Eq. (3.7). Since in the workspace $x_4 < x_2$, the solution with a negative sign yields larger y.

3.4.2 Inverse Kinematics

Parallel manipulators frequently have an inverse kinematics problem that is simpler than the direct kinematics problem. The Pantograph is no exception. The problem is to find the angles θ_1 and θ_5 given the position of point $\mathbf{P}_3 = (x_3, y_3)$. A pentagon

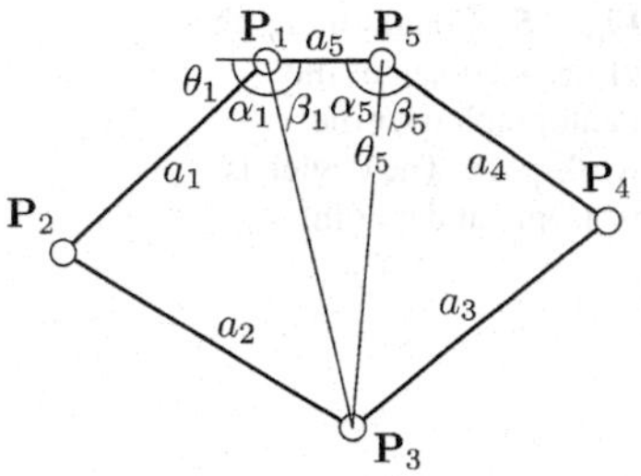

Fig. 3.4 Dividing the pentagon into three triangles

can be divided into three triangles, see Fig. 3.4 which makes the solution straightforward:

$$\theta_1 = \pi - \alpha_1 - \beta_1, \qquad \theta_5 = \alpha_2 + \beta_2, \tag{3.8}$$

where

$$\alpha_1 = \arccos\left(\frac{a_1^2 - a_2^2 + \|\mathbf{P}_1 - \mathbf{P}_3\|^2}{2a_1\sqrt{\|\mathbf{P}_1 - \mathbf{P}_3\|}}\right), \tag{3.9}$$

$$\beta_1 = \operatorname{atan2}(y_3, -x_3), \tag{3.10}$$

$$\beta_5 = \arccos\left(\frac{a_4^2 - a_3^2 + \|\mathbf{P}_5 - \mathbf{P}_3\|^2}{2a_4\sqrt{\|\mathbf{P}_5 - \mathbf{P}_3\|}}\right), \tag{3.11}$$

$$\alpha_5 = \operatorname{atan2}(y_3, x_3 + a_5). \tag{3.12}$$

This solves the inverse kinematics for a generic Pantograph with arm lengths a_i, as long as the device is in a configuration such that $\alpha_1 > 0$ and $\beta_5 > 0$, which puts it in the permitted workspace.

3.4.3 Differential Kinematics

The Jacobian matrix can be found by direct differentiation of the direct kinematic map with respect to the actuated joints θ_1 and θ_5:

$$\mathbf{J} = \begin{bmatrix} \partial x_3/\partial\theta_1 & \partial x_3/\partial\theta_5 \\ \partial y_3/\partial\theta_1 & \partial y_3/\partial\theta_5 \end{bmatrix} = \begin{bmatrix} \partial_1 x_3 & \partial_5 x_3 \\ \partial_1 y_3 & \partial_5 y_3 \end{bmatrix}, \tag{3.13}$$

where $\partial_i \cdot$ denotes the partial derivative with respect to θ_i. Let $d = \|\mathbf{P}_2 - \mathbf{P}_4\|$, $b = \|\mathbf{P}_2 - \mathbf{P}_h\|$ and $h = \|\mathbf{P}_3 - \mathbf{P}_h\|$.

Applying the chain rule to Eqs. (3.6) and (3.7):

$$\partial_1 x_2 = a_1 \sin(\theta_1), \qquad \partial_1 y_2 = a_1 \cos(\theta_1), \tag{3.14}$$

$$\partial_5 x_4 = a_4 \sin(\theta_5), \qquad \partial_5 y_4 = a_4 \cos(\theta_5), \tag{3.15}$$

$$\partial_1 y_4 = \partial_1 x_4 = \partial_5 y_2 = \partial_5 x_2 = 0, \qquad \partial_i h = -b\partial_i b/h, \tag{3.16}$$

$$\partial_i d = \frac{(x_4 - x_2)(\partial_i x_4 - \partial_i x_2) + (y_4 - y_2)(\partial_i y_4 - \partial_i y_2)}{d}, \tag{3.17}$$

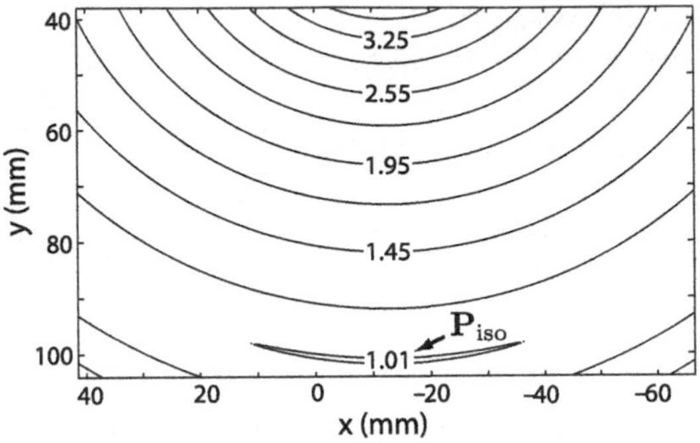

Fig. 3.5 Condition number of the Jacobian of the Pantograph over the workspace. The device is isotropic at the point $\mathbf{P}_{\text{iso}}$

$$\partial_i b = \partial_i d - \frac{\partial_i d(a_2^2 - a_3^2 + d^2)}{2d^2}, \tag{3.18}$$

$$\partial_i y_h = \partial_i y_2 + \frac{\partial_i bd - \partial_i db}{d^2}(y_4 - y_2) + \frac{b}{d}(\partial_i y_4 - \partial_i y_2), \tag{3.19}$$

$$\partial_i x_h = \partial_i x_2 + \frac{\partial_i bd - \partial_i db}{d^2}(x_4 - x_2) + \frac{b}{d}(\partial_i x_4 - \partial_i x_2), \tag{3.20}$$

$$\partial_i y_3 = \partial_i y_h - \frac{h}{d}(\partial_i x_4 - \partial_i x_2) - \frac{\partial_i hd - \partial_i dh}{d^2}(x_4 - x_2), \tag{3.21}$$

$$\partial_i x_3 = \partial_i x_h + \frac{h}{d}(\partial_i y_4 - \partial_i y_2) + \frac{\partial_i hd - \partial_i dh}{d^2}(y_4 - y_2). \tag{3.22}$$

3.4.4 Kinematic Conditioning

All entries of the Jacobian have the dimension of lengths mapping angular velocities $\omega = [\dot{\theta}_1\ \dot{\theta}_5]^\top$ to linear velocities $v = [\dot{x}_3\ \dot{y}_3]^\top$: $v = \mathbf{J}\,\omega$. Thus, the 2-norm of the Jacobian matrix (which also is a length) has the physical meaning of scaling the sensor nominal resolution to the nominal resolution of the device. The Jacobian matrix is well conditioned on all the workspace and the device becomes isotropic at (Fig. 3.5):

$$\theta_{1\,\text{iso}} = \arccos\left(-\frac{25}{126} + \frac{25\sqrt{2}}{42}\right), \qquad \theta_{5\,\text{iso}} = \pi - \theta_{1\,\text{iso}} \tag{3.23}$$

corresponding to the point $\mathbf{P}_{\text{iso}} \simeq (-12.5,\ 101.2)$ in Fig. 3.5. At this point the two distal links intersect orthogonally at the tip and the end effector is equidistant from the actuated joints. Here, the Jacobian matrix maps disks in the angular velocity joint space to disks in the tip velocity space. There are just two such points. The other point which has a negative y is not used. The isotropic region is near the edge of the workspace but this is an acceptable compromise given that the main objective is dynamic performance. The device, as dimensioned, has a large region of dynamic near-isotropy spreading over most of the workspace [13].

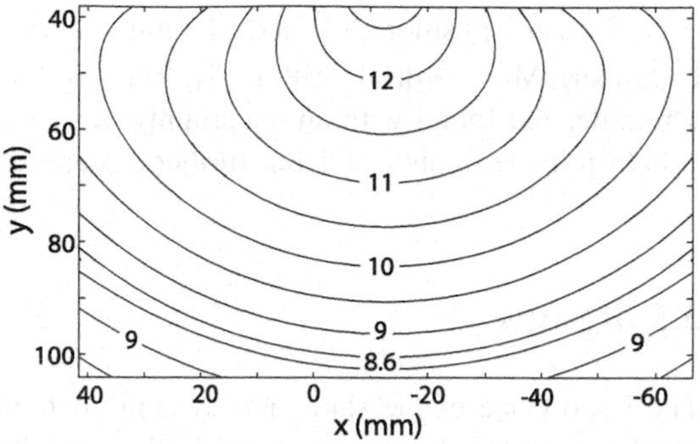

Fig. 3.6 Resolution of the Pantograph in the workspace, measurement unit is the µm. The device is equipped with two encoders with 2^{16} counts per revolution, the resolution is $\|\Delta\mathbf{X}\| = \|\mathbf{J}\|_2\sqrt{2}\frac{2\pi}{2^{16}}$

If $\|\cdot\|_2$ denotes the largest singular value of a matrix, then expression:

$$\|\Delta\mathbf{X}\| \le \|\mathbf{J}\|_2 \, \|[\Delta\theta_1 \, \Delta\theta_5]^\top\|, \tag{3.24}$$

where $\Delta\mathbf{X} = [\Delta x \; \Delta y]^\top$ is the resolution of the device and $\Delta\theta_i$ the resolution of an encoder. This allows us to plot the ideal resolution of the device in Fig. 3.6 for the case where encoders with 65K CPR are used.

3.4.5 Calibration

Since the angles are measured by incremental encoders, the origin needs to be calibrated at system startup. The workspace of the device is mechanically limited to a rectangular area which can be used for this purpose. In a first maneuver, point $\mathbf{P}_3$ is brought by the user to the bottom left corner of the workspace to roughly calibrate the encoders. The user then proceeds to acquire many calibration points by sliding the end effector along the four edges (bottom, right, top, left). Points acquired on the bottom edge all have the same y coordinate, so on this edge, $\mathbf{P}_3^{i\downarrow} = (x_3^i, y^\downarrow)$ where $y^\downarrow$ is the known common value of the coordinate, and similarly for the other edges: $y^\uparrow$ for the top edge, $x^\leftarrow$ for the left edge, and $x^\rightarrow$ for the right.

Call the θ_1^i and θ_5^i, the measurements acquired and x_3 and y_3 the components of the direct kinematic function: $\mathbf{P}_3 = [x_3(\theta_1, \theta_5) \; y_3(\theta_1, \theta_5)]^\top$. The device can be calibrated by minimizing the error function

$$E = \sum_{i=1}^{N_\downarrow} (y^\downarrow - y_3(\theta_1^{i\downarrow} + \theta_1^0, \theta_5^{i\downarrow} + \theta_5^0))^2 \tag{3.25}$$

$$+ \sum_{i=1}^{N_\rightarrow} (x^\rightarrow - x_3(\theta_1^{i\rightarrow} + \theta_1^0, \theta_5^{i\rightarrow} + \theta_5^0))^2 \tag{3.26}$$

$$+ \sum_{i=1}^{N_\uparrow} (y^\uparrow - y_3(\theta_1^{i\uparrow} + \theta_1^0, \theta_5^{i\uparrow} + \theta_5^0))^2 \tag{3.27}$$

$$+ \sum_{i=1}^{N_\leftarrow} (x^\leftarrow - x_3(\theta_1^{i\leftarrow} + \theta_1^0, \theta_5^{i\leftarrow} + \theta_5^0))^2, \tag{3.28}$$

over the zero positions θ_1^0 and θ_5^0: $\min_{\theta_1^0,\theta_5^0} E$. This is accomplished using the Levenberg-Marquardt algorithm [8]. The results are satisfying since the two offset angles are found with an uncertainty of 6–7 counts which can be attributed to backlash in the joints 2 and 4 as further discussed in Sect. 3.5.2.

3.5 Results

The importance of the static and dynamic behavior of haptic devices, accounting for the mechanical structure, transmission and drive electronics has been well recognized by device designers [1, 2, 7, 14, 19, 23].

Guidelines for measuring the performance characteristics of force feedback haptic devices were documented in [12]. Among these guidelines two are particularly important, in addition to the usual requirement of minimizing interference with the process being measured. The first specifies that the characteristics must be measured where the device is in contact with the skin. The second recognizes the fact that a haptic device has a response that depends on the load. Therefore, load reflecting the conditions of actual use must be applied during the measurements. From this view point, measurement of the system response from the actuator side and without a load, as it is sometimes done (e.g., [4]), fails to provide the sought information. A useful actuator-side technique that quantifies the structural properties of a device in terms of a "structural deformation ratio" (SRD) was nevertheless suggested [18]. It was not used here since the complete system response provides richer information.

3.5.1 Experimental System Response

The frequency response (from amplifier current command to acceleration at the tip) was measured with a system analyzer (DSP Technology Inc., SigLab model 20-22) using chirp excitation. This technique was used because it is more precise and more robust to nonlinearities (and more time consuming) than an ARMAX procedure.

Measurements were performed under three conditions. The first corresponded to the unloaded condition. In order to prevent the device from drifting away during identification, it was held in place by a loosely taught rubber band. The second condition was created by lightly touching the interface plate while the response was measured. In the third condition, the device was loaded by pressing firmly on it.

An ideal device should have a uniform gain across all frequencies (and would have to have a SDR index of 1.0 [18]). Figure 3.7 shows all three responses on the same graph but offset by 10 dB for clarity. The response is indeed flat over a wide bandwidth (40 to 300 Hz). But accidents occur in the low and the high frequency regions.

In the low frequency region, the rise in gain for the "unloaded response" was most probably due to presence of the rubber band and can be ignored. However contact

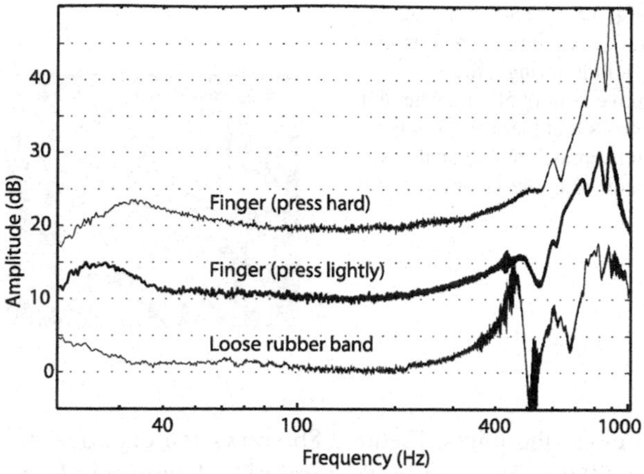

Fig. 3.7 Frequency Response of the device when an identical signal is sent to both the amplifiers to create an horizontal movement. The intensity of the movement is measured with an accelerometer approximately parallel to the movement. The response curves relative to the finger are shifted of +10 dB (light pressure) and +20 dB (hard pressure)

with a finger creates a low Q resonance (5 dB or less) which shifted up in frequency when the finger pressed harder. This could be explained by the non-linear nature of tissues. These observations conspire to indicate that indeed, it would be difficult to reduce the finger dynamics to that of a linear time invariant system without risking to oversimplify the dynamics of the actual system [10, 17].

In the high frequency region, there were two notable events in the response. The "unloaded response" first shows what is the typical fingerprint of a sharp, low-loss structural resonance (pole-zero pair) in the 400–500 Hz band. This could be attributed to flexibility *inside* the motor as these often emit acoustic noise at this frequency upon torque transients (this is also the case of all haptic interfaces using the same "bell coreless" motors). As the finger presses harder on the interface, this resonance is progressively masked by the load but probably continues to occur, but is unseen at the tip. Now, what is more difficult to explain are additional structural accidents in the 900 Hz region, which instead of being attenuated by a larger load as one would expect, are actually enhanced to reach up to 30 dB of gain, a rather large magnitude indeed. If these were due to structural resonance of the linkages, then one would observe a shift in frequency due to non-linear buckling. But it is not the case. This problem will be further discussed in Sect. 3.6. In the meantime we establish that the device can reliably be used in the DC–400 Hz range provided that proper roll-off filters are used.

3.5.2 Resolution

We estimated the actual device resolution using the setup shown in Fig. 3.8a. A micropositioner was connected to joint 3 along so it could back-drive the device along the y axis in the vicinity of point $\mathbf{P}_{iso}$. Backlash and other joint imperfections were likely to deteriorate the resolution of the device but should not be considered first. To minimize their influence, a constant torque was applied by the motor to

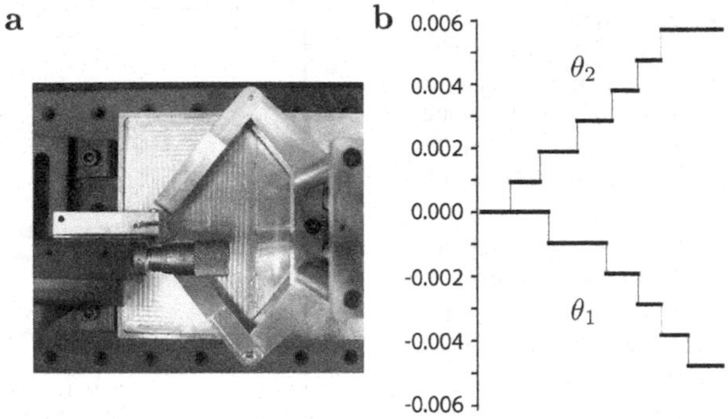

Fig. 3.8 (**a**) Setup used to verify resolution. (**b**) Encoder reading during a linear movement of 50 μm. The plot shows that there are 5 or 6 ticks, matching the analysis made with the Jacobian

preload the joints. Figure 3.8b shows the encoders values when the tip is moved by 50 μm. This verifies the resolution determined from the analysis of the Jacobian matrix.

3.6 Conclusion and Discussion

This paper has described the redesign of the Pantograph haptic device with the aim to increase its performance so it would be capable of providing high quality haptic rendering. Its performance was evaluated and found to meet the initial expectation of uniform and high bandwidth response. However, while the device operates very well, several points still need attention. The manner in which they can be addressed is now discussed by order of increasing implementation difficulty.

1. The backlash in the joints in certain conditions, particularly when the plate is not statically loaded, can reach several encoders ticks. The cause was simple to find and so will be the solution to eliminate it. The present bearings were specified of ordinary quality. In fact, their backlash specifications match the observations. They should be replaced by higher quality bearings since clearly this is a limiting factor.
2. The device structure is machined out of aluminum. It is possible that the metallic structure participates in the observed unwanted high frequency resonance. Composite materials could be used to manufactured haptic devices with structural properties designed to optimize their response (e.g., adjust for critical damping) [16].
3. The device should incorporate a source of calibrated viscous damping [5], something which is the subject of on-going work.
4. "Bell coreless" motors work well but are less than ideal for haptic device applications due to (1) their sharp internal resonance characterized in this paper, and (2) use of un-needed brushes in a limited angle application [24]. Motors having an absence of torque ripple, absence of cyclical reluctant torque (cogging), optimized structural properties, and absence of friction (in addition to high torque, of course) should be designed specifically for this application. Recent proposals for

electronic compensation of the injurious properties of motors designed for other purposes fall short of our requirements in this respect [15].

Finally, we hope to be able to release the system publicly in a near future, even if not all the points discussed above are fully addressed. At the present time however, the system is in use to carry out studies in high fidelity friction and texture synthesis techniques [3, 11].

Acknowledgements Qi Wang and Gianni Campion thank PRECARN Inc. for scholarships. The authors would also like to thank Hsin-Yun Yao for assistance in PCB design and manufacturing and Andrew Havens Gosline for insightful comments on an earlier draft of this paper.

References

1. Adelstein, B.D., Rosen, M.J.: Design and implementation of a force reflecting manipulandum for manual control research. In: Proceedings of the ASME Dynamic Systems and Control Division, vol. 42, pp. 1–12 (1992)
2. Buttolo, P., Hannaford, B.: Pen-based force display for precision manipulation in virtual environments. In: Proceedings of Virtual Reality Annual International Symposium, pp. 217–224 (1995)
3. Campion, G., Hayward, V.: Fundamental limits in the rendering of virtual haptic textures. In: Proceedings of the First Joint Eurohaptics Conference and Symposium on Haptic Interfaces for Virtual Environments and Teleoperator Systems, WHC'05, pp. 263–270 (2005)
4. Cavusoglu, M.C., Feygin, D., Tendick, F.: A critical study of the mechanical and electrical properties of the PHANToM haptic interface and improvements for high performance control. Presence **11**(6), 555–568 (2002)
5. Colgate, J.E., Schenkel, G.G.: Passivity of a class of sampled-data systems: Application to haptic interfaces. J. Robot. Syst. **14**(1), 37–47 (1997)
6. DiFilippo, D., Pai, D.K.: Contact interaction with integrated audio and haptics. In: Proceedings of the International Conference on Auditory Display, ICAD (2000)
7. Ellis, R.E., Ismaeil, O.M., Lipsett, M.: Design and evaluation of a high-performance prototype planar haptic interface. In: Proc. ASME Advances in Robotics, Mechatronics, and Haptic Interfaces, vol. DSC-9, pp. 55–64 (1993)
8. Fletcher, R.: Practical Methods of Optimization. Wiley, New York (1987)
9. Grant, D.: Two new commercial haptic rotary controllers. In: Proceedings of the First Joint Eurohaptics Conference and Symposium on Haptic Interfaces for Virtual Environments and Teleoperator Systems, WHC'05 (2004)
10. Hasser, C.J., Cutkosky, M.R.: System identification of the human hand grasping a haptic knob. In: Proceedings of the 10th Symposium on Haptic Interfaces for Virtual Environment and Teleoperator Systems, pp. 171–180 (2002)
11. Hayward, V., Armstrong, B.: A new computational model of friction applied to haptic rendering. In: Corke, P., Trevelyan, J. (eds.) Experimental Robotics VI. Lecture Notes in Control and Information Sciences, vol. 250, pp. 403–412 (2000)
12. Hayward, V., Astley, O.R.: Performance measures for haptic interfaces. In: Giralt, G., Hirzinger, G. (eds.) Robotics Research: The 7th International Symposium, pp. 195–207. Springer, Heidelberg (1996)
13. Hayward, V., Choksi, J., Lanvin, G., Ramstein, C.: Design and multi-objective optimization of a linkage for a haptic interface. In: Lenarcic, J., Ravani, B. (eds.) Advances in Robot Kinematics, pp. 352–359. Kluver Academic, Dordrecht (1994)

14. Hayward, V., Gregorio, P., Astley, O., Greenish, S., Doyon, M., Lessard, L., McDougall, J., Sinclair, I., Boelen, S., Chen, X., Demers, J.-P., Poulin, J., Benguigui, I., Almey, N., Makuc, B., Zhang, X.: Freedom-7: A high fidelity seven axis haptic device with application to surgical training. In: Casals, A., de Almeida, A.T. (eds.) Experimental Robotics V. Lecture Notes in Control and Information Science, vol. 232, pp. 445–456. Springer, Berlin (1998)
15. Lawrence, D.A., Pao, L.Y., White, A.C., Xu, W.: Low cost actuator and sensor for high-fidelity haptic interfaces. In: Proc. 12th International Symposium on Haptic Interfaces for Virtual Environment and Teleoperator Systems, HAPTICS'04, pp. 74–81 (2004)
16. McDougal, J., Lessard, L.B., Hayward, V.: Applications of advanced materials to robotic design: The freedom-7 haptic hand controller. In: Proceedings of the Eleventh International Conference on Composite Materials, ICCM-11 (1997)
17. Milner, T.E., Franklin, D.W.: Characterization of multijoint finger stiffness: Dependence on finger posture and force direction. IEEE Trans. Biomed. Eng. **45**(11), 1363–1375 (1998)
18. Moreyra, M., Hannaford, B.: A practical measure of dynamic response of haptic devices. In: Proceedings of the IEEE International Conference on Robotics and Automation, pp. 369–374 (1998)
19. Morrell, J.B., Salisbury, J.K.: Performance measurements for robotic actuators. In: Proceedings of the ASME Dynamic Systems and Control Division, vol. 58, pp. 531–537 (1996)
20. Murayama, J., Bougrila, L., Luo, Y., Akahane, K., Hasegawa, S., Hirsbrunner, B., Sato, M.: SPIDAR G&G: A two-handed haptic interface for bimanual VR interaction. In: Proceedings of EuroHaptics 2004, pp. 138–146 (2004)
21. Quanser: Haptic Devices. http://www.quanser.com/
22. Ramstein, C., Hayward, V.: The pantograph: A large workspace haptic device for a multimodal human-computer interaction. In: Proceedings of the SIGCHI Conference on Human Factors in Computing Systems, CHI'04, ACM/SIGCHI Companion-4/94, pp. 57–58 (1994)
23. Rosenberg, L.: How to assess the quality of force-feedback systems. J. Med. Virtual Real. **1**(1), 12–15 (1995)
24. Salcudean, S.E., Stocco, L.: Isotropy and actuator optimization in haptic interface design. In: Robotics and Automation, 2000. Proceedings. ICRA '00. IEEE International Conference on, vol. 1, pp. 763–769 (2000)

Chapter 4
Fundamental Limits

Abstract We discuss the properties of force feedback haptic simulation systems that fundamentally limit the re-creation of periodic gratings, and hence, of any texture. These include sampling rate, device resolution, and structural dynamics. Basic sampling limitations are analyzed in terms of the Nyquist and the Courant conditions. The analysis proposes that noise due to sampling and other sources injected in the system may prevent it to achieve acceptable performance in most operating conditions, unless special precautions such as the use of a reconstruction filter, make the closed-loop more robust to noise. The structural response of a PHANTOM 1.0A device was such that no such filter could be found, and the system introduced heavy distortion in gratings as coarse as 10 mm. The Pantograph Mark-II device having more favorable structural properties could reliably create gratings between 1 and 10 mm.

4.1 Preface to Chap. 4

This chapter introduces a novel framework for assessing the rendering capabilities of force feedback haptic devices. Six necessary conditions are proposed for the precise rendering of virtual haptic textures. Two force feedback haptic devices are characterized according to the new theory; then, they are tested over a spatially varying texture-like force field. The results confirm the fidelity of textures rendered with the improved Pantograph haptic device, thus validating the design and the new control introduced in the previous chapter. Moreover, one of the conditions introduced in this chapter is related to the passivity of the texture rendering algorithm used to generate the force field. This condition is further developed and studied in two chapters of this book, because it is a powerful analysis tool for the quality of haptic texture algorithms.

This chapter lays part of the theoretical foundation of the work presented in this book; and the most direct consequence of this work is the validation of the capa-

Reprinted from Gianni Campion and Vincent Hayward, "Fundamental Limits in The Rendering of Virtual Haptic Textures." *Proc. First Joint Eurohaptics Conference and Symposium on Haptic Interfaces for Virtual Environments and Teleoperator Systems WHC'05*, 2005, pp. 263–270.

G. Campion, *The Synthesis of Three Dimensional Haptic Textures: Geometry, Control, and Psychophysics*, Springer Series on Touch and Haptic Systems,
DOI 10.1007/978-0-85729-576-7_4, © IEEE 2005

bilities of the Pantograph, which is then used in the rest of the book whenever an experimental setup for haptic textures is required.

4.1.1 Contributions of Authors

Gianni Campion characterized both haptic devices, developed and implemented the experimental tests, and derived the conditions and authored the paper. Prof. Hayward contributed to the derivation of the conditions, supervised the experimental and identification process, and edited the manuscript and figures. The authors acknowledge the contribution of Hsin-Yun Yao, who designed and fabricated the accelerometer used in the identification phase.

4.2 Introduction

Texture is important in haptic simulations because, like frictional properties or shape, it is a key attribute of real and simulated objects. In computer graphics, much work was, and still is, aimed at texturing images, but in haptics, despite much past research [6, 7, 9, 11, 16–19], the question of realism has only been recently addressed [4].

In this paper, we discuss the characteristics of a system which set absolute limits on what can be rendered with force feedback devices. Because these devices operate on sampled data both in time and in space, artifacts can arise when a user interacts with virtual objects using a mechanical interface which necessarily interposes its own dynamics between the object and the user's fingers.

In general, the factors that limit the synthesis of texture independently from any particular method are: system sampling period, sensor noise (related to resolution), output torque resolution, device structural dynamics, and other factors such as backlash in the joints.

What we found is that for commonly available devices, the finest textures that can be reliably and accurately synthesized without special precautions are surprisingly and perilously coarse.

4.3 Basic Sampling

All texture synthesis algorithms rely on a "generating function" $g(x)$ used to compute a force from a position. It can be periodic, stochastic, or a mixture of both. To analyze sampling effects, we must assume that $g(x)$ is C^1 given that it must be finitely sampled and then reconstructed during synthesis. Without loss of generality, we also assume $|g(x)| \leq 1, \forall x$, and that it is band-limited. For this reason, it is sufficient to look at the case of a sinusoidal grating. The observations made next

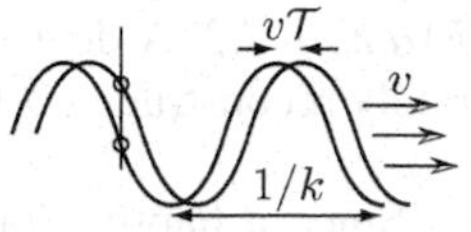

Fig. 4.1 *Circles* represent places where the grating is ideally sampled

extend to any periodic grating and to band-limited stochastic textures since they can be decomposed in a finite sum of sinusoids.

Consider scanning a grating of spatial period $1/k$ at an unknown velocity v. The system is sampled at rate $1/\mathcal{T}$. This is equivalent to sampling a progressive wave with wave number k at a stationary point, see Fig. 4.1.

There exists a critical velocity v_T at which the discretized grating vanishes,

$$v_T = \frac{1}{k\mathcal{T}}. \tag{4.1}$$

Precise reconstruction is possible only if $v \ll v_T$, i.e. $v < \alpha v_T$. The Nyquist criterion states that to reconstruct a signal, we need $\alpha \geq 2$. If this condition is not met then the generating function cannot be reconstructed. The accuracy of the reconstruction, however, is not guaranteed by the Nyquist criterion. We must pick a safer limit, since the Nyquist rate can be approached only when using near-ideal reconstruction filters. Since, typically, the electromechanical transfer function of the haptic device serves as a less-than-ideal reconstruction filter, $\alpha \approx 10$ provides a reasonable limit. In this condition,

$$g(x(t)) = \sin(2\pi k x(t)) \tag{4.2}$$

can be well approximated by its discrete-time counterpart

$$g_i(x_i) = \sin(2\pi k x_i), \tag{4.3}$$

where the x_i are the successive position samples measured at rate $1/\mathcal{T}$ by the device. When $v \ll v_T$, then the temporal frequency is $f = vk \ll 1/\mathcal{T}$ so the system operates far from the Nyquist rate, that is:

$$\boxed{\alpha k v \mathcal{T} < 1} \tag{4.4}$$

Example 4.1 Simulate a grating with a 1.0 mm pitch at 1.0 kHz ($k = 10^3$, $\mathcal{T} = 10^{-3}$). Staying sufficiently far under v_T requires the scanning speed to remain under 0.1 m/s—a rather low speed by human standards. Simulating a finer pitch of 0.1 mm (grooves of a vinyl record) would require the speed to remain under 0.01 m/s.

So far we assumed that the device measured the x_i perfectly. In practice, any device has limited resolution. Let's consider that the device makes quantized measurements with a resolution δ, the smallest displacement at the tip that can be reliably detected. By an analogous reasoning, in the absence of a space reconstruction filter, then would require at least $\beta \approx 10$ samples within one spatial period:

$$\boxed{\beta k \delta < 1} \tag{4.5}$$

Example 4.2 A device having a resolution of 10 μm ($\delta = 10^{-5}$) at best can accurately reconstruct a 0.1 mm grating ($k = 10^{-4}$), all other sources of error ignored.

Since a haptic simulation system essentially solves a numerical problem discretized in time and space, it is subject to the Courant-Friedrichs-Lewy condition:

$$\delta > v_C \mathcal{T}, \quad v_C < \frac{\delta}{\mathcal{T}}, \tag{4.6}$$

which is equivalent to considering that the velocity cannot be known better than one velocity quantum $\delta/\mathcal{T}$. We can conclude that increasing the sampling rate of the simulation may not improve it, if the resolution of the device is not increased as well, while making it more difficult to estimate velocity [5, 14].

Example 4.3 If we sample at 10.0 kHz ($\mathcal{T} = 10^{-4}$), for the device to resolve movements at $v = 0.1$ m/s, its resolution must be better than 10 μm ($\delta = 10^{-5}$).

Therefore, a trade-off exists between device resolution, sampling rate, scanning speed and grating period. We can reconcile these observations by combining the safe velocity of Eq. (4.4) given by the Nyquist criterion with the critical velocity of Eq. (4.6) given by the Courant condition ($\alpha = v\mathcal{T}/\delta$ is the Courant number) and find that, indeed, the device should have a resolution such that

$$\boxed{\alpha k \delta < 1} \tag{4.7}$$

Output quantization can also cause similar problems. If we call b the smallest step of force that can be resolved, we should impose that there are at least γ steps within the rendered force amplitude A:

$$\boxed{\gamma b < A} \tag{4.8}$$

These constraints can easily be extended to non-periodic textures by knowing the spectrum of the generating function.

An estimate of how well the generating function can be reconstructed is found by considering that each measurement is made with an error on δ_i on the true position $x_i = x + \delta_i$. Assuming that δ_i is small:

$$g(x + \delta_i) \approx g(x) + \partial g/\partial x\, \delta_i = g(x) + \epsilon \tag{4.9}$$

$$\approx \sin(2\pi k x) + 2\pi \delta_i k \cos(2\pi k x). \tag{4.10}$$

Thus, the discrete grating has an error term ϵ:

$$|\epsilon|_\infty = 2\pi k \max_i |\delta_i| = 2\pi \delta k. \tag{4.11}$$

For a given device resolution the error is amplified if the spatial period is smaller. This error is dominated by space quantization when v is small, $\epsilon \simeq 2\pi/\beta$ and by time quantization when v is large: $\epsilon \simeq 2\pi/\alpha$ (Eqs. (4.5) and (4.7)).

Example 4.4 Simulate a 1.0 mm grating with a device with a resolution of 10 μm ($\delta = 10^{-5}, k = 10^3$, $\beta = 100$). The relative error is 0.06, that is 6%. If we try to simulate a finer grating of 0.1 mm pitch ($\beta = 10$), the error become 60% which is hardly acceptable. Since δ is random, the simulation is noisy.

4.4 Feedback Dynamics

While many results were found in the past by considering a device to be a damped mass (a rigid body plus some dissipation), when it comes to simulating textures, it is clear that device behavior at high frequencies matters, and that the rigid body assumption may no longer hold.

A texture simulation system operates in closed loop, hence, feedback control theory can be useful to analyze its properties. Consider the classical set-up as in Fig. 4.2 [1, 8]. There, r represents an input to be tracked, y the output, d the noise injected in the system normalized at the output, e the error, and often one adds an external disturbance n to the nominal command u giving v. The loop is closed around a controller C and a plant P.

In texture simulation, sources for d (encoder noise) and n (numerical noise and other disturbances such as friction or analog-to-digital reconstruction noise) have been identified in the preceding section. Two of these transfer functions are of particular importance, that of d to y (transmission function T) and that of d to v. Calling $L = PC$,

$$T \triangleq -\frac{y}{d} = \frac{L}{1+L}, \qquad \frac{v}{d} = \frac{T}{P} = \frac{C}{1+L}. \tag{4.12}$$

Similar manipulations would allow us to evaluate the effect of n on the closed-loop system.

We now relate the general diagram to the case of a haptic texture simulation system as in Fig. 4.3. In most instances, actuators and sensors are co-located, so P represents the device transfer function from motor current command to motor movement (including the amplifiers).

While the closed-loop function T is crucial, what ultimately matters is what the user feels, thus the system response should be considered from the device tip (measured with an accelerometer [10]). This corresponds to an open-loop transfer function R which is related to the displacement h of skin via a double integrator.

The functions R and P co-vary as a function of many factors: as a function of the configuration, of the load, and in particular, from the mere act of touching the device. This is because devices naturally have structural characteristics with high-Q resonances (e.g., [3, 13]), possibly also arising from the motors [2]. Structural dynamics are uncertain and resonances can shift unpredictably. A systematic design for a filter H would be difficult (μ-synthesis, convex optimization, or other methods), and if at all possible, will have to be conservative. The structural dynamics of a device, both amplify noise in the close-loop, and distort the signal in open loop. Since it is hard to robustly compensate for structural dynamics beyond the first mode, this introduces another fundamental limit. Calling F_0, the frequency of the first mode of the device:

$$\boxed{vk < F_0} \tag{4.13}$$

The generating function yields a force signal of maximum amplitude A, the intensity of the resulting grating. Linearizing g around a particular true position x as in Eq. (4.10) (small signal analysis), gives us a block C ($r = 0$) corresponding to the

Fig. 4.2 System with feedback

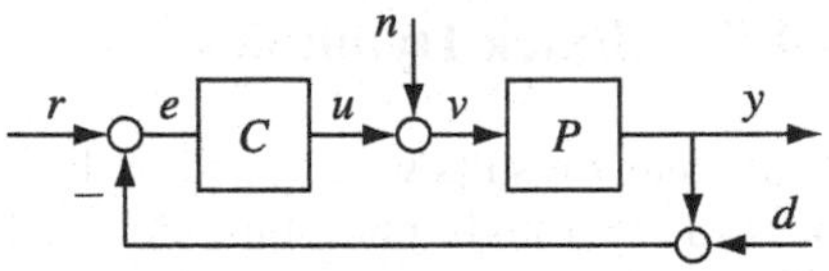

Fig. 4.3 Haptic simulation set-up

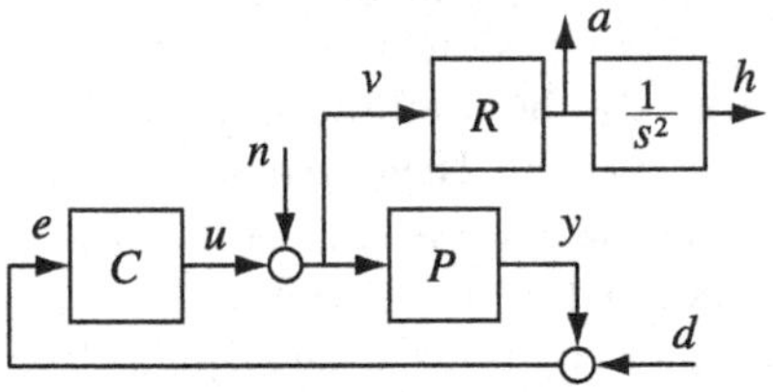

slope of the texture generating function times the intensity factor A (the Jacobian of g in the case of multidimensional texture simulation) plus a reconstruction filter H that is typically ignored (i.e. $H = 1$):

$$C = A\frac{\partial g}{\partial x}H. \tag{4.14}$$

The finer the texture, the higher the instantaneous loop gain, which varies with k for the simple grating (Eq. (4.10)). This introduces a new constraint that says that in order to keep the loop gain independent from any particular grating, then A must be reduced proportionally to k. Calling A_0 the maximum acceptable stiffness (e.g., for stability):

$$\boxed{2\pi A k < A_0} \tag{4.15}$$

4.5 Experiments

We applied the foregoing analysis using two haptic devices: the PHANTOM® from Sensable and the recently re-built Pantograph Mk-II [2], see Fig. 4.4. The PHANTOM (model 1.0A) is a haptic device designed to explore 3D objects which is frequently used in research laboratories. It has cable drives that provide torque amplification and is statically balanced [15]. The Pantograph is a direct-driven planar device designed to render surfaces [12].

4.5.1 Device Characterization

Sampling rate The devices were both interfaced to a personal computer (2.5 GHz P-IV processor), via a PCI proprietary interface for the PHANTOM, and via a "hardware-in-the-loop" PCI card from Quanser Inc. (Model Q8) for the Pantograph. The system was running RTLinux 3.2pre3 that enabled hard real-time sampling rates

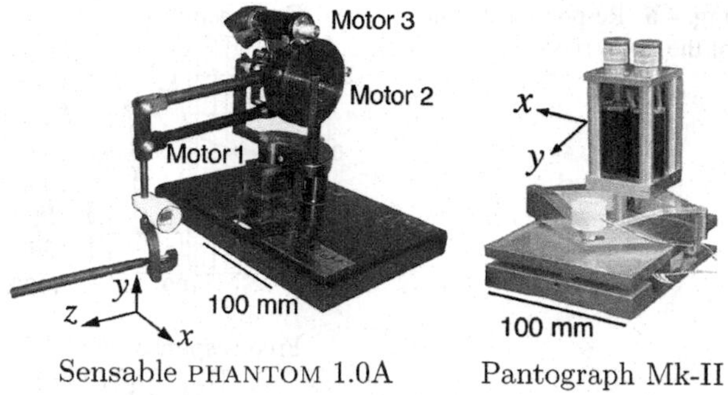

Fig. 4.4 Devices used for the experiment

up to 100 kHz. In all cases, however, the control loops ran at 10 kHz. We found that at 10 kHz, ($\mathcal{T} = 10^{-4}$) RTLinux ran the hard-realtime thread with a period jitter never exceeding 0.5%.

Resolution Given Δ the vector of p individual joint resolutions and $\mathbf{J}(q)$ the device Jacobian, the resolution was estimated using:

$$|\delta|_\infty = \max_{\mathbf{l}\in\{-1,1\}^p} (\|\mathbf{J}(q)\,\mathrm{diag}(\Delta)\mathbf{l}\|). \tag{4.16}$$

For the PHANTOM 1.0A (4,000 CPR encoders and accounting for joint ratios), the nominal resolution was found to vary between 40 and 70 μm, Fig. 4.5. For the Pantograph Mk-II with 2^{16} CPR encoders, the nominal resolution was found to vary between 9 and 13 μm, see Fig. 4.5. Note that, in essence, these figures express resolutions *which are guaranteed not to be achieved in practice*.

Nevertheless, the safety factor β found in Sect. 4.3 is greater than 10 for both devices for textures whose smallest spatial period is 1 mm. Given 12 bit analog-to-digital converters, the PHANTOM's force granularity varied between 13 and 20 mN and the Pantograph's between 2 and 5 mN.

Structural Response The devices were tested using chirp excitation (DSP Technology Inc. system analyzer, SigLab Model 20-22). An accelerometer (Analog Device; model ADXL250) was clamped to the distal end of the PHANTOM using a

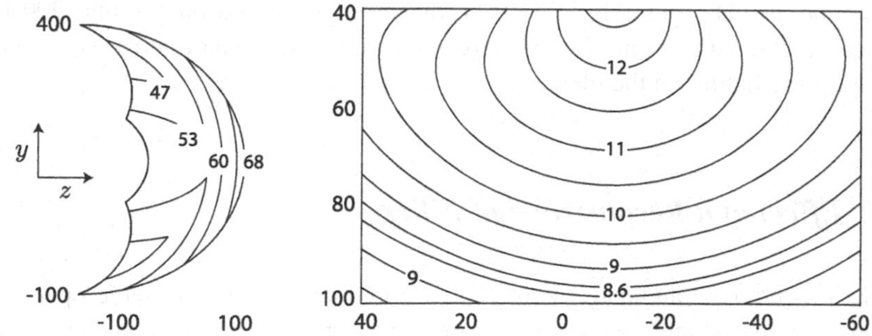

Fig. 4.5 The nominal resolution in μm of the PHANTOM in the mid-sagittal plane (*left*) and of the Pantograph (*right*) plotted over their workspaces indicated in millimeters

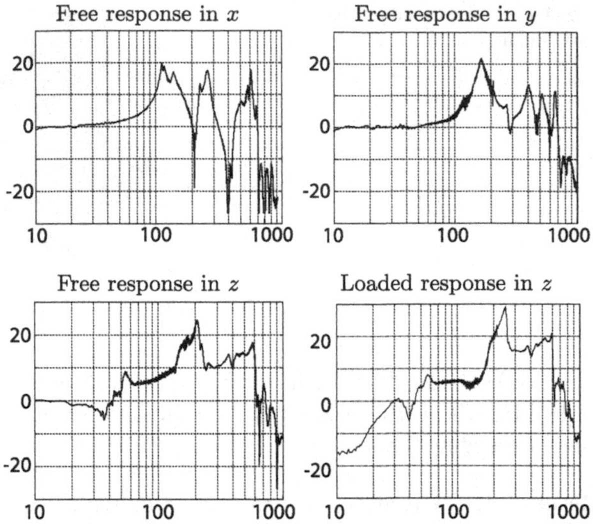

Fig. 4.6 Response examples of the PHANTOM

light-weight fixture to minimize effects on the response. For the Pantograph, the same accelerometer was embedded in the finger interface plate. This enabled us to measure directly the open-loop transfer function R. We did not attempt to measure P since encoders do not have enough resolution in the high frequencies where displacements are vanishingly small.

For the PHANTOM, the condition where the device was lightly loaded by a rubber-band (slightly taught to keep it in place) is reported for all directions. It is also shown for the z direction when it was loaded by a grip.

The results, Fig. 4.6, indicate that the lowest structural anti-resonance was around 30 Hz in the z direction and that there was a resonance at 100 Hz on the x axis. Naturally, there were many others modes extending up to 700 Hz, changing in frequency, magnitude, and Q, according to the loading conditions.

For the Pantograph, the first condition also was when the plate was held with by a rubber-band, the second was when a finger touched the plate lightly, and the third when the finger pressed hard, see Fig. 4.7. There were two dominant resonances, one around 400 Hz (possibly introduced by the motors) and one around 900 Hz. The second resonance is puzzling because it magnified instead of being damped out when pushing harder on the plate.

4.5.2 Effect of a Reconstruction Filter

For the PHANTOM, candidate filters H that could provide a reasonable open loop response while making the closed loop more robust given the structural imperfections at 30 Hz and 100 Hz were found to introduce too much phase delay, making the system unstable.

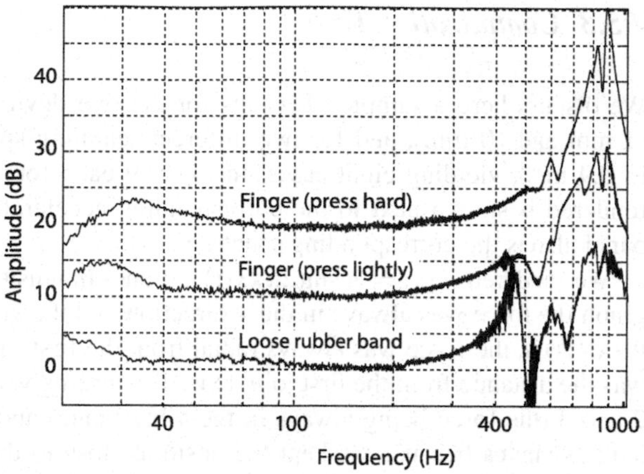

Fig. 4.7 Dynamic response of the Pantograph. The three responses are shown offset by 10 dB for clarity

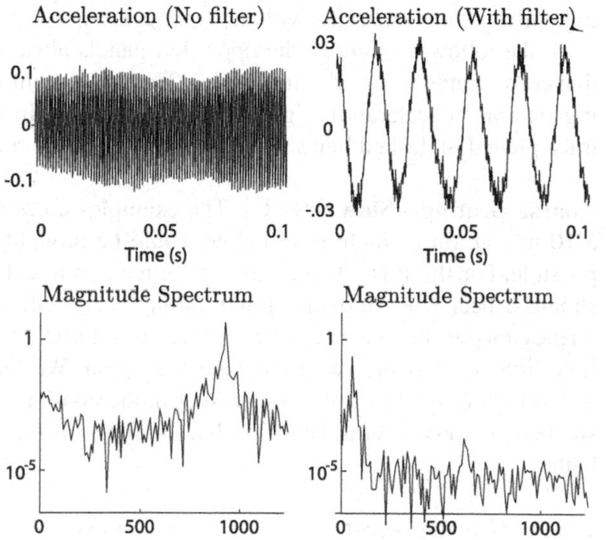

Fig. 4.8 Pantograph: Effect of 400 Hz Butterworth filter: *Left* without filter, *right* with the filter in the loop

For the Pantograph, because the response was well behaved until 400 Hz, a filter could be empirically designed (Butterworth order 10, 400 Hz cut-off, ran at 10 kHz). With this filter, the Pantograph could in principle render a 1 mm grating with an error of 8% and a speed of 0.4 m/s.

Figure 4.8 shows the effect of adding the filter when exploring a 1 mm sinusoidal grating. Without it, the rendered texture is essentially uncorrelated with the $g(x)$. The left panels show that the rendered acceleration is essentially determined by the noise injected in the system. It is magnified and almost exclusively concentrated in the 900 Hz band, as shown by the acceleration spectrum. With the filter, the rendered texture, see the right panels, has the expected shape and most of its energy is in the correct frequency band.

4.5.3 Comparative Tests

We discuss here a sample of results for the two devices, for two grating periods (1 mm and 10 mm), and for two different scanning speed ranges (from 0.06 m/s to 0.9 m/s), yielding eight cases. In Fig. 4.9 each force waveform that was to be rendered is shown next to the corresponding acceleration waveform. Under each panel also is the corresponding spectrum.

We rendered $f(s) = A\sin(2\pi ks(t))$, a one-dimensional grating. For the Pantograph the force was always in the x direction and the $s(t)$ was simply $x(t)$. For the PHANTOM, the force was always radial from the first joint and horizontal, and $s(t)$ was the distance from the first joint axis. The grating was rendered with just motors 2 and 3 (the force being always in the 4-bar plane) and depended mostly on the z axis dynamics because we kept the position close to the neutral point. The values for A where 0.9 N for the PHANTOM and 0.4 N for the Pantograph, providing a conservative γ margin. These values also provided similar tactile intensities as well as similar stability margins with the two devices.

In the following figures, the upper-left panels always show the force command that corresponds to signal u in Fig. 4.3. The upper-right panel shows to corresponding measured acceleration, signal $a = s^2h$ in Fig. 4.3. In the upper panels, the second trace plotted in dashed line shows acceleration in an orthogonal direction.

Coarse Grating—Slow Speed The examples corresponded to slowly scanning a 10 mm grating which, in principle, could be thought of providing the best case possible. For the PHANTOM, while the force command was not sinusoidal, as one should expect when moving slowly through an oscillatory force field (in this frequency range, the elasticity of the finger is significant), the resulting acceleration bore little resemblance with the expected signal. We see that the PHANTOM introduced high frequency noise not present in the commanded signal. The Pantograph showed the same overall behavior, but the shape of the acceleration signal is much better.

Coarse Grating—Fast Speed These examples used the same grating, but this time, the experimenter attempted to move at a relatively fast speed. As expected, the force command signal was then confined to a narrow band because the movement 'punches through' the grating, and this was the case for both devices. For the PHANTOM, the rendering was acceptable in the direction of the movement as shown by the magnitude spectrum of the acceleration. There was a noticeable defect in the orthogonal direction where, somehow, significant energy spills over in the 700 Hz range. The observed spectrum spread is the hallmark of a non-linear system. The cross-talk in orthogonal directions also changes the signal frequency. For the Pantograph, the rendering was nearly perfect.

Fine Grating—Slow Speed These examples corresponded to the case of a 1 mm grating scanned at slow speed. The force to be rendered by the PHANTOM suffers from quantization noise (even if $\beta > 10$) and some instability in the 300 Hz range.

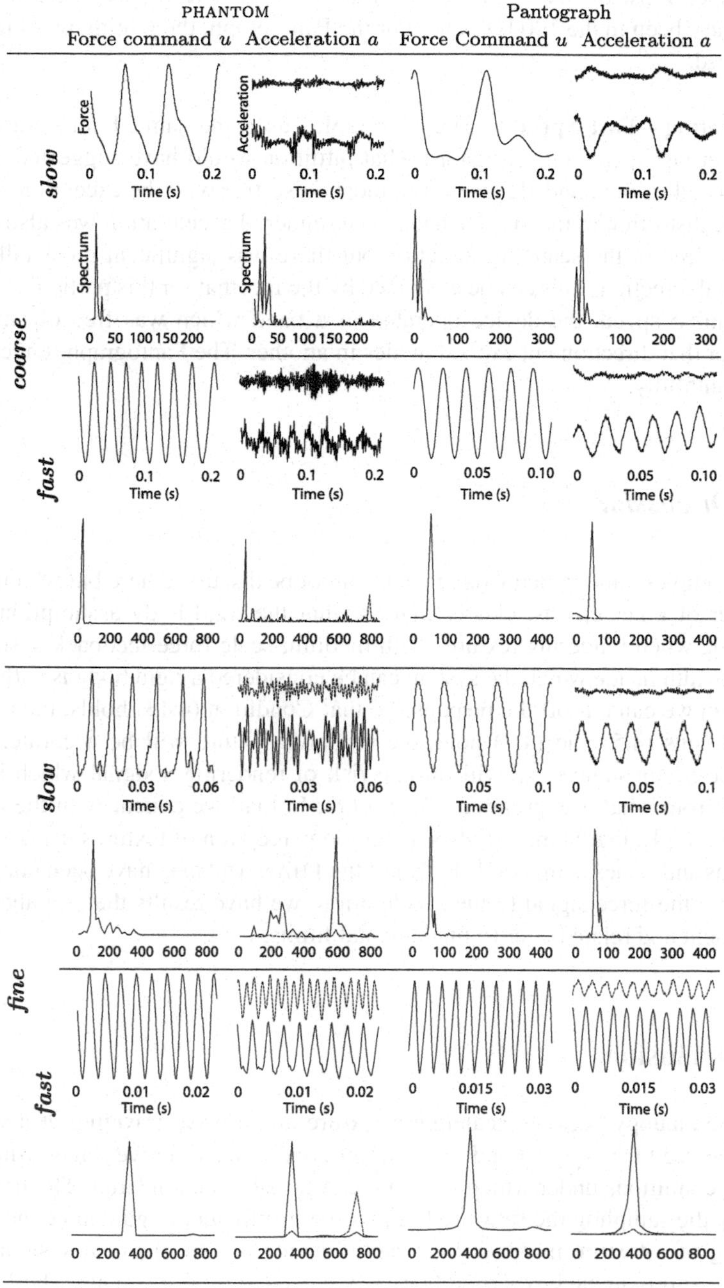

Fig. 4.9 Summary of the eight testing conditions for the two devices

This yielded an acceleration signal which was essentially unrelated to the desired result, since most of the signal energy was in the 600 Hz band where in fact it should have been in the 100 Hz band. For the Pantograph, the grating was faithfully reproduced.

Fine Grating—Fast Speed These examples used the same 1 mm grating but scanned at fast speed. Going against what intuition would have suggested, for the PHANTOM, the commanded force was almost noise-free with the exception of some harmonic distortion in the 700 Hz band. The rendered acceleration was also almost distortion-free in the scanning direction but there was significant cross-talk in an orthogonal direction. This can be explained by the fact that for this particular grating and scanning speed, the device operated in a band which was free of structural modes for that direction but excited modes in another. The Pantograph rendered the grating faithfully.

4.5.4 Discussion

These examples, among many others that cannot be discussed here but which reveal a number of other effects, clearly indicate that the rigid body assumption is not acceptable when rendering textures with medium-scale force feedback systems. If the bandwidth inside which the system can be considered a rigid body is sufficiently large, then we can cut-off the response so that Condition (4.13) holds, then as long as Conditions (4.5) and (4.7) hold, the rendered texture will be accurate. Under any other circumstances we will run the risk of rendering a signal which is quite different from what was programmed in $g(x)$. In that, we concur with the opinion expressed in [4], that human studies about the perception of textures using systems of designs and scales comparable to that of the PHANTOM may have been tainted. By comparing the force signal to the acceleration, we have results that are algorithm-independent and relate better to fundamental limits.

4.6 Conclusion

Using the analogy between scanning a texture and a wave traveling at a variable speed, we used the Nyquist and the Courant conditions to derive relationships that state the conditions under which a texture can possibly be rendered. The limits imposed by the sampling theory were found to be insufficient to guarantee the correct rendering of a texture in general. A haptic device is a mechanical system which cannot be approximated by a rigid body when excited by fast signals. The Jacobian of the rendering function essentially determines the gain in the closed loop, therefore the complete system is subject to the constraints of feedback dynamics when significant noise is injected in a system which is structurally non-robust.

Table 4.1 Summary of limits

Scanning velocity limit	$\alpha k v T < 1$
Low speed reconstruction limit	$\beta k \delta < 1$
High speed reconstruction limit	$\alpha k \delta < 1$
Force reconstruction limit	$\gamma b < A$
Gain limit	$2\pi A k < A_0$
Device structural limit	$v k < F_0$

What we found can be summarized as follows. Given k the spatial frequency of a grating, $\mathcal{T}$ the system sampling period, v the scanning velocity, δ the device resolution, b the force resolution, α a temporal safety factor (at least 2, most likely 10), β a spatial safety factor (at least 2, most likely 10 or more), γ a force reconstruction safety factor (at least 10), A the desired force amplitude the rendered grating, A_0 the maximum control stiffness, and F_0 the first mode of the device, then Table 4.1 summarizes the limits that cannot be exceeded in order to make it possible to render a given grating with a given device. These limits do not guarantee that the grating question will be rendered correctly, but if one of these limits is exceeded it is highly likely that it will not be the case.

As an example, the PHANTOM which, in principle, has enough resolution in time and space to render correctly textures up to 1 mm was found to render incorrectly textures as coarse as 10 mm. With another device, the Pantograph, which has a much higher structural bandwidth, it was possible to find a reconstruction filter which robustified the system under all reasonable operating conditions, although finding optimal filters that can take into account both the open loop and the closed loop behavior of a given haptic system remains a daunting task.

This study also suggested a new performance measure for haptic devices, namely the smallest grating that can be rendered reliably. For the PHANTOM, we were not able to determine it. For the Pantograph, this number is around 1 mm, still a far cry indeed from what is needed to simulate a realistic texture imitating surface finishes, such as that of wood, for example.

Acknowledgements This research was supported in part by the Institute for Robotics and Intelligent Systems, and the Natural Sciences and Engineering Research Council of Canada. G. Campion is the recipient of a PRECARN Inc. scholarship.

The authors would like to thank Prof. Hong Z. Tan of Purdue University for insightful comments on an early draft of this paper and the reviewers for their excellent suggestions. The authors are indebted to Prof. David Ostry of McGill University for letting us use his laboratory's PHANTOM, to Prof. Keyvan Hashtrudi-Zaad of Queen's University for showing us how to interface it; to Hsin-Yun Yao of the Haptics Lab at McGill for custom-packaging the miniature accelerometers, and to Andrew Havens Gosline also from the Haptics Lab for proof-reading the paper.

The authors would like to acknowledge Seigo Harashima from RICOH Company for many keen discussions on haptic textures.

References

1. Barratt, C., Boyd, S.: Closed-loop convex formulation of classical and singular value loop shaping. In: Leondes, C.T. (ed.) Digital and Numeric Techniques and Their Applications in Control Systems, Part 1 (1993)
2. Campion, G., Wang, Q., Hayward, V.: The Pantograph Mk-II: A haptic instrument. In: Proceedings of the IEEE/RSJ International Conference on Intelligent Robots and Systems, IROS'05, pp. 723–728 (2005)
3. Cavusoglu, M.C., Feygin, D., Tendick, F.: A critical study of the mechanical and electrical properties of the PHANToM haptic interface and improvements for high performance control. Presence **11**(6), 555–568 (2002)
4. Choi, S., Tan, H.Z.: Perceived instability of virtual haptic texture. I. Experimental studies. Presence **13**(4), 395–415 (2004)
5. Colgate, J.E., Brown, J.M.: Factors affecting the Z-width of a haptic display. In: Proceedings of the IEEE International Conference on Robotics and Automation, pp. 3205–3210 (1994)
6. Costa, M.A., Cutkosky, M.R.: Roughness perception of haptically displayed fractal surfaces. In: Proceedings ASME IMECE Symposium on Haptic Interfaces for Virtual Environments and Teleoperator Systems, vol. 69-2, pp. 1073–1079 (2000)
7. Crossan, A., Williamson, J., Murray-Smith, R.: Haptic granular synthesis: Targeting, visualisation and texturing. In: Proceedings of the International Symposium on Non-visual & Multimodal Visualization, pp. 527–532. IEEE Press, New York (2004)
8. Doyle, S.P., Francis, B.A., Tannenbaum, A.R.: Feedback Control Theory. Macmillan, New York (1992)
9. Fritz, J.P., Barner, K.E.: Stochastic models for haptic textures. In: Stein, M.R. (ed.) Telemanipulator and Telepresence Technologies III. Proc. SPIE, vol. 2901, pp. 34–44 (1996)
10. Hayward, V., Astley, O.R.: Performance measures for haptic interfaces. In: Giralt, G., Hirzinger, G. (eds.) Robotics Research: The 7th International Symposium, pp. 195–207. Springer, Heidelberg (1996)
11. Hayward, V., Yi, D.: Change of height: An approach to the haptic display of shape and texture without surface normal. In: Siciliano, B., Dario, P. (eds.) Experimental Robotics VIII. Springer Tracts in Advanced Robotics, pp. 570–579. Springer, Heidelberg (2003)
12. Hayward, V., Choksi, J., Lanvin, G., Ramstein, C.: Design and multi-objective optimization of a linkage for a haptic interface. In: Lenarcic, J., Ravani, B. (eds.) Advances in Robot Kinematics, pp. 352–359. Kluver Academic, Dordrecht (1994)
13. Hayward, V., Gregorio, P., Astley, O., Greenish, S., Doyon, M., Lessard, L., McDougall, J., Sinclair, I., Boelen, S., Chen, X., Demers, J.-P., Poulin, J., Benguigui, I., Almey, N., Makuc, B., Zhang, X.: Freedom-7: A high fidelity seven axis haptic device with application to surgical training. In: Casals, A., de Almeida, A.T. (eds.) Experimental Robotics V. Lecture Notes in Control and Information Science, vol. 232, pp. 445–456. Springer, Berlin (1998)
14. Janabi-Sharifi, F., Hayward, V., Chen, C.-S.J.: Discrete-time adaptive windowing for velocity estimation. IEEE Trans. Control Syst. Technol. **8**(6), 1003–1009 (2000)
15. Massie, T.H., Salisbury, J.K.: The PHANToM haptic interface: A device for probing virtual objects. In: Proceedings ASME IMECE Symposium on Haptic Interfaces for Virtual Environments and Teleoperator Systems, vol. DSC-Vol. 55-1, pp. 295–301 (1994)
16. Minsky, M., Lederman, S.J.: Simulated haptic textures: Roughness. In: Proceedings of the ASME IMECE Symposium on Haptic Interfaces for Virtual Environments and Teleoperator Systems, vol. DSC-Vol. 58, pp. 421–426 (1996)
17. Otaduy, M.A., Lin, M.C.: A perceptually-inspired force model for haptic texture rendering. In: Proceedings of the 1st Symposium on Applied Perception in Graphics and Visualization, pp. 123–126. ACM Press, New York (2004)
18. Siira, J., Pai, D.K.: Haptic textures—a stochastic approach. In: Proceedings of IEEE International Conference on Robotics and Automation, pp. 557–562 (1996)
19. Weisenberger, J.M., Kreier, M.J., Rinker, M.A.: Judging the orientation of sinusoidal and square-wave virtual gratings presented via 2-DOF and 3-DOF haptic interfaces. Haptics-e **1**(4) (2000), online

Chapter 5
On the Synthesis of Haptic Textures

Abstract Advanced, synthetic haptic virtual environments require textured virtual surfaces. We found that texturing smooth surfaces often reduces the system passivity margin of a haptic simulation. As a result, a smooth virtual surface that can be rendered in a passive manner may loose this property once textured. We propose that any texture algorithm is associated with a characteristic number that expresses the relative change in loop gain. We further found that a passive virtual interaction can have severe unwanted artifacts if the synthesized force field is not conservative. The energy characteristics of seven algorithms are analyzed. Finally a new texture synthesis algorithm, which operates by modulating a friction force during scanning, is shown to have several advantages over previous ones.

5.1 Preface to Chap. 5

This chapter develops one of the conditions found in Chap. 4, and introduces the novel concept of characteristic number of a texture algorithm. In addition, the energy properties of several commonly used texture algorithms are reported and related to an artifact usually known as "aliveness". Finally, a novel formulation of a friction based texture rendering algorithm is proposed.

The most important contribution of this chapter is the definition of the characteristic number for planar surfaces: when a texture is added to a flat virtual wall, the change in passivity margin is algorithm specific and can be summarized by the characteristic number. This number can be used to analyze the effect of different texture parameters over the quality of the haptic rendering.

The characteristic number is used in the rest of the book to evaluate the passivity of the texture algorithms used in the psychophysical experiments.

5.1.1 Contributions of Authors

Gianni Campion worked on the mathematical aspects of the paper, implemented the texture algorithms, collected and analyzed the data, produced the figures, and wrote

Reprinted from Gianni Campion and Vincent Hayward, "On the Synthesis of Haptic Textures." *IEEE Transactions on Robotics*, Volume 24, Number 3, 527–536, 2008.

G. Campion, *The Synthesis of Three Dimensional Haptic Textures: Geometry, Control, and Psychophysics*, Springer Series on Touch and Haptic Systems, DOI 10.1007/978-0-85729-576-7_5, © IEEE 2008

the manuscript of the paper. Prof. Hayward closely followed the development of the paper, advised on the math and on the experimental process, edited the manuscript and the figures.

The authors identified an inaccuracy in the paper after its publication. The last appendix to this chapter contains an erratum to the paper and was published: Erratum to "On the synthesis of haptic textures" 2009, IEE Transactions on Robotics, vol. 25, issue 2, page 475.

5.2 Introduction

The "virtual wall" is considered to be a benchmark problem in haptic simulation [1, 2, 9, 12, 30]. Nevertheless, advanced, high-fidelity haptic simulations demand to consider more general cases than smooth, low curvature virtual object boundaries that virtual walls can represent.

This article considers the question of augmenting the simulation of rigid or deformable objects with surface texture. The aim is to provide a method to systematically analyze the properties of a given texture simulation algorithm, since many algorithms have been proposed [10, 11, 14, 17, 21, 25, 26, 28, 29, 32, 34], and many are yet to be designed.

An approach to designing haptic synthesis algorithms for simulating basic mechanical interactions between objects is to consider algorithms that provide fundamental properties shared with the interactions that these algorithms are supposed to simulate. Those properties should include general physical properties such as mathematical continuity of the force response and conservation of energy. It could be further desired to provide resemblance of the synthetic force responses with actual physical responses [23]. Passivity is also an important property of synthetic environments because, once provided, the simulation results no longer depend on the dynamic properties of the hand interacting with it [24].

In this paper, it is found that texture synthesis algorithms frequently increase the *effective stiffness* of an originally non-textured object. This algorithm-specific gain, if not accounted for, can lead to the loss of passivity during a virtual haptic interaction. A characteristic number is introduced to express the relative increase of stiffness caused by a given algorithm when texture is added to a smooth virtual surface. In addition, the cause for unintended artifacts, such as low frequency oscillations [6, 7], observed when using certain algorithms is elucidated and attributed to the creation of non-conservative virtual force fields. A noteworthy consequence is that even if a synthesis method yields a passive interaction, it can produce simulation artifacts that severely impair the realism of a simulation by causing virtual objects to feel active [31].

5.3 Assumptions

5.3.1 Parametrization

Most texturing algorithms, see Sect. 5.5, assume the existence of a bounded height function h defined on a smooth surface. For analysis purposes it is convenient to

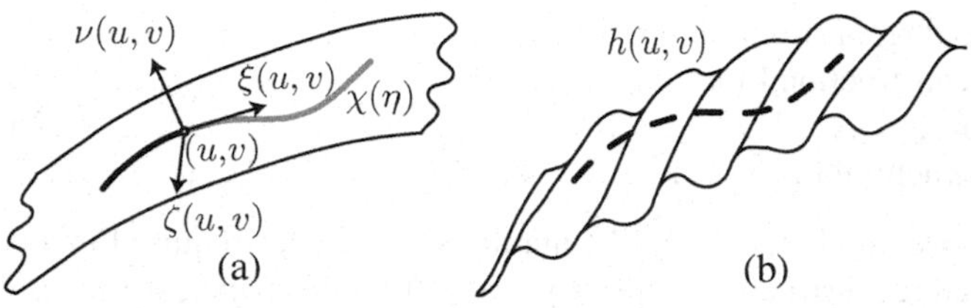

Fig. 5.1 Parametrization of a tool interacting with a textured surface

consider that each scan path defines a curve parametrized by its arc length, $\chi(\eta)$, which is located on the smooth surface, Fig 5.1a. This path also defines a moving Frenet frame located at a point on the surface parametrized in (u, v). This frame provides local coordinates $\xi(u, v)$, $\nu(u, v)$, and $\zeta(u, v)$, the tangent, normal, and binormal vectors respectively. A height function, $h(u, v)$, can then be naturally defined by the distance measured between the textured surface and the smooth boundary along the normal $\nu(u, v)$.

Intuitively, if a surface has low curvature, the effects of texture dominate. We therefore restrict the present study to surfaces with zero curvature and straight paths. Specializing the analysis to the cross plane x–z, then suffice to study the case of $z = h(x, 0)$ when h is defined on the surface $z = 0$ and when the scan is along x. With suitable changes of coordinates, it is possible to extend the present analysis to account for the effects of surfaces with high curvature or for the effects of blending functions along the edges of polygonal meshes. These extensions are left to future work.

The continuity and differentiability requirements of h are discussed later in this article in greater detail, but in general we assume that while $h(x, y)$ can be periodic, stochastic, or otherwise, it must be band-limited so the Nyquist-Shannon sampling theorem can apply. We also consider that the haptic device has a limited bandwidth, which can be assumed to be determined by its first mechanical resonance F_0. This is a good assumption because it is difficult to robustly compensate beyond the first mode in closed loop [35].

5.3.2 Limits

A set of necessary, yet not sufficient conditions must be met for a haptic texture to be properly synthesized [3]. Given the assumptions just stated, the analysis can be restricted to the case of a sinusoidal texture with spatial period l and amplitude A. Calling $\mathcal{T}$ the system sampling period, v the scanning velocity of the manipulandum, δ the spatial resolution of the device, b the force resolution of the device, A the maximum amplitude of the force to be displayed, and A_0 the maximum stiffness achievable by the device, there are six inequalities which must be met:

1. Scanning velocity limit .... $\alpha v \mathcal{T}/l < 1$
2. Low speed reconstruction limit .... $\beta\delta/l < 1$

3. High speed reconstruction limit .... $\alpha\delta/l < 1$
4. Force reconstruction limit .... $\gamma b/l < A$
5. Gain limit .... $2\pi A/l < A_0$
6. Device structural limit .... $v/l < F_0$

In the above, α, β, and γ are the numbers of samples required in a period for a correct causal reconstruction (typically 5 to 10). In this paper, we focus on the fifth inequality and assume that the five others are met. The notation is this paper is by and large consistent with that in [18] and [3], yet here, stiffnesses are denoted by 'κ' to avoid conflict with singular values noted σ.

5.4 Control Analysis

5.4.1 Control Passivity Condition

We first derive a corollary of Theorem 1 in reference [24], which is a generalization of Colgate and Schenkel's passivity condition [8]. Our corollary states that the largest singular value of the Jacobian matrix of the force field describing the synthetic force response must be smaller than the smallest singular value of the damping matrix of the device divided by twice the sampling period,

Corollary 5.1 *Let* $s(\eta) = [x(\eta), y(\eta), z(\eta)]^\top$ *be a trajectory of the virtual tool, which is in general different from* $\chi(\eta)$. *A texturing algorithm generates a force trajectory in the static field* $\boldsymbol{f}(x, y, z)$. *Call* $\mathbf{B}$ *the device damping matrix with smallest singular value* $\sigma_0(\mathbf{B})$, *and* $\mathbf{J}_f$ *the Jacobian matrix of* $\boldsymbol{f}(x, y, z)$ *with largest singular value* $\sigma_n(\mathbf{J}_f)$. *Given time steps i*

$$\forall i, \quad \|\mathbf{J}_{f,i}\|_2 = \sigma_n(\mathbf{J}_{f,i}) \leq \frac{\sigma_0(\mathbf{B})}{2T}, \tag{5.1}$$

is a sufficient condition for the passivity of the virtual environment.

Proof Adapt Condition 1 of Theorem 1 of [24]:

$$|\boldsymbol{f}(s_i) - \boldsymbol{f}(s_{i-1})| \leq \frac{\sigma_0(\mathbf{B})}{2T}|s_i - s_{i-1}|. \tag{5.2}$$

To a first order approximation, if $\mathbf{J}_f$ is the Jacobian matrix of $\boldsymbol{f}(s)$, then

$$|\mathbf{J}_{f,i}[s_i - s_{i-1}]| \leq \|\mathbf{J}_{f,i}\|_2|s_i - s_{i-1}|, \tag{5.3}$$

therefore if $\|\mathbf{J}_{f,i}\| \leq \sigma_0(B)/2T$ then

$$|\mathbf{J}_{f,i}[s_i - s_{i-1}]| \leq \frac{\sigma_0(\mathbf{B})}{2T}|s_i - s_{i-1}| \tag{5.4}$$

which proves the Corollary by applying the triangular inequality to (5.4) and (5.3). □

In short, this Corollary means that the algorithm used to synthesize texture has an effect on the passivity of the virtual object. For memoryless algorithms, the effect can be quantified independently of the path chosen to explore the surface. For algorithms with memory, a worst-case analysis can be performed. It is a generalization of Condition 5 of Sect. 5.3.

5.4.2 Characteristic Number of Algorithms

Under the assumption that the vector field f is linear in the stiffness of the underlying smooth wall κ_0, we define

$$q = \frac{\sigma_n(\mathbf{J}_f)}{\kappa_0} \tag{5.5}$$

to be a characteristic number associated with the algorithm generating f. This number expresses a stiffness increase when simulating a "texturized" boundary. Suppose for example that a device is capable of rendering a smooth virtual wall of stiffness κ_0. Once textured, in order to preserve passivity the original wall stiffness should be reduced to $\kappa_1 = \kappa_0/q$. This notion applies to any texturing algorithm studied in this article. The determination of q requires finding the maximum singular value of $\mathbf{J}_f$ inside the regions where the device is intended to produce textures.

The knowledge of the characteristic numbers is useful to design virtual environments. For example, it will enable a designer to predict that doubling the spatial frequency of a texture requires dividing the nominal stiffness by two. It must be noted however, that it cannot be used as a tool to directly compare algorithms since its value can depend on different sets of parameters.

5.4.3 Conservativity and Passivity in Virtual Environments

Physically, a system is said to be conservative if the work done to modify it is equal to the change of its internal energy. Equivalently, if work is done to modify the system through *any* cycle, the system is conservative if, and only if, this work is zero. In classical mechanics, barring friction and internal dissipation, all forces are conservative [15]. Conservativity is therefore a desirable property of any virtual environment intended to replicate objects which are not actuated, as originally indicated by Salisbury et al. [31]. Later, we will use two equivalent properties to study the algorithms: the gradient of a scalar function is a conservative vector field and if a vector field has zero curl then it is conservative.

Passivity, in turn, is typically defined with reference to input-output properties in a system [22]. In this view, a system is seen as a black box and dissipativity is defined in an abstract sense in terms of an input-output product (e.g., forces and displacements for mechanical systems). A system must always extract energy

Table 5.1 Expressions for **f(s)**. See Fig. 5.2 for the corresponding diagrams and main text for details

A	$[0, -\kappa_0 d]^\top$ [6]	**B & G**	$-\kappa_0 d\boldsymbol{n}$ [6] & [27]
D	$-\kappa_0 d^{\min}\hat{\boldsymbol{n}}$ [36]	**E**	$[0, -\kappa_0 d^{\min}]^\top$
F	$[-\kappa(x)d, -\kappa_0 d]^\top$	**C**	$\left[-\kappa_0 \frac{\partial h}{\partial p^x} d, -\kappa_0 d\right]^\top$ [17]

through time to be passive. If we combine a conservative system with dissipative elements, such as friction or dampers, seen from outside the box, the system will necessarily be passive. Nevertheless, if we combine a generative system with elements that are sufficiently dissipative, the system, as a whole, may remain passive.

For the design of *multi-dimensional* virtual mechanical environments [24], conservativity is therefore a more desirable property than input-output passivity since one could construct systems that behave passively for some trajectories but have severely non-physical responses for others. A useful consequence of enforcing conservativity for a virtual environment is to neatly decouple the performance of a synthesis algorithm from that of a device and from its control. Three cases can arise:

1. Condition (5.1) does not hold, then unwanted energy may be injected in the system.
2. Condition (5.1) holds and the algorithm is generative, then the simulation may inject energy for some trajectories even if the simulation is passive for other trajectories.
3. Condition (5.1) holds and the virtual environment is conservative, then for any trajectory the system will always extract energy from the user.

Unless the simulation is intended to be generative, the third case is the only desirable case.

5.5 Analysis of Algorithms

We now investigate seven different algorithms for their energy properties in x–z coordinates only (see Sect. 5.3). Figure 5.2 and Table 5.1 summarize the algorithms, all sharing a nominal stiffness κ_0.

Algorithms can be classified into three groups according to how penetration is computed.

- Algorithms **A**, **B**, and **C** determine d by projection $d = d^z = p^z - h(p^x)$, where p is the position of the interaction point in virtual space.
- Algorithms **E** and **D** use the minimum distance between p and the textured surface $d = d^{\min}$ as a measure of penetration.
- Algorithms **F** and **G** compute d as the distance to the nominal surface, not to the textured surface: $d = p^z$.

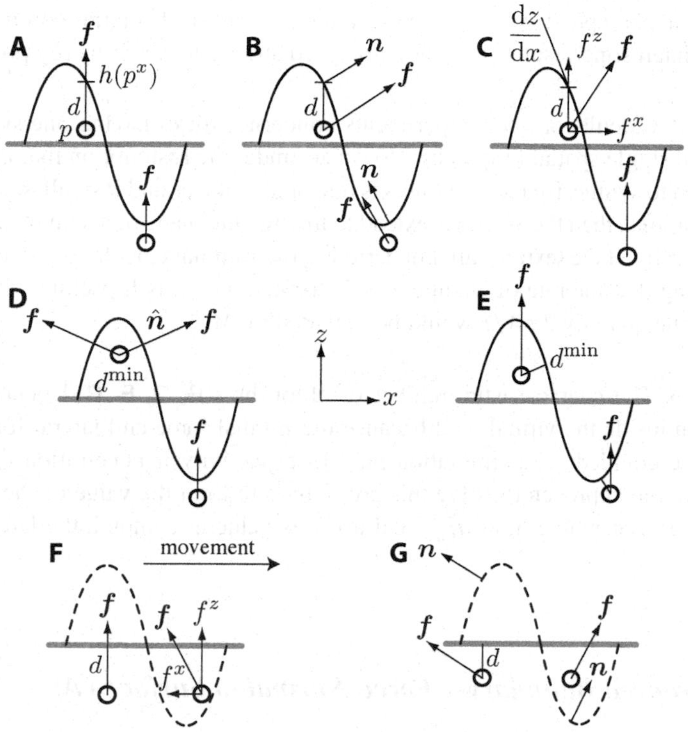

Fig. 5.2 Different texture synthesis algorithms. *Circles* represent the position of the handle in virtual space. For each algorithm we show the force in two different locations. *Thick gray lines* symbolize the non-textured surface. *Solid lines* indicate the virtual object boundary that is used in the penetration computation. *Dashed lines* represent the textured surfaces when the penetration is computed from the nominal surface. **A**, **B** are the algorithms used in [6]. **C** is discussed in [17]. **E** and **D** are derived from the 'god object' approach [36]. **F** is a dry-friction-modulation based algorithm introduced here. **G** is from [27]

Remark 5.1 It is important to recognize that an accurate reproduction of haptic texture is a difficult problem, not only from the point of the view of the device performance limitations [20], but also from the view point of the physics involved. There is a variety of complex micro-mechanical phenomena related to the tribological properties of the materials in contact and to the mutual geometrical relationships of the tool and the surface. It is a common experience that the texture of rough paper is greatly affected by the instrument used to write on it. Compare a pencil and a ball-pen. For the same paper, the micro-mechanics and the feel are different. In one case, the interaction depends on graphite dry friction and on the other, on a ball rolling on viscous fluid bed. Moreover, a given texture synthesized with a given algorithm often produces different perceptual experiences when felt through haptic devices of different types. In view of these complexities, the algorithms considered in this paper are neither discussed from the view point of their physical relevance nor from that of their perceptual value—all have plausible physical interpretations and all may give interesting results. They are discussed strictly for their energy properties.

Remark 5.2 Calculations and experiments concern textures having one single amplitude and single spatial frequency. However, under the assumption that the haptic device used to synthesize these textures is reasonably linear under small-signal operating conditions, then these results extend to any texture made of a sum of sinusoids. More generally, if the textures are not periodic, then similar calculations can be done by assuming that the rate of change of h is limited, i.e., it is Lipschitz with bound M. Then, the quantity $2\pi A/l$ would be replaced by M.

Remark 5.3 The characteristic number of algorithms **B**, **G**, **F**, **C** depends on the penetration inside the virtual wall because the normal force and lateral force components are coupled. The simulation may lose passivity if penetration exceeds a limit. A simple approach to solve this problem is to limit the value of the penetration d^z to an acceptable limit $d^z_{\max}$ and use this value to compute the lateral force component.

5.5.1 *Grooved Boundary—Force Normal to Surface* (**A**)

5.5.1.1 Field

Algorithm **A** considers a virtual-wall-like force component that is normal to the nominal non-textured surface, [6]. There is no force component along the direction tangent to the nominal virtual surface. The texture sensation then arises by virtue of the oscillations caused by a vector field aligned along one single direction.

$$f_{\mathsf{A}}(s) = \begin{cases} [0, -\kappa_0 d^z]^\top & \text{if } d^z < 0, \\ [0, 0]^\top & \text{otherwise.} \end{cases} \tag{5.6}$$

The penetration is computed from the boundary of the texture along the z direction.

5.5.1.2 Jacobian

Inside the boundary, when $d < 0$, the Jacobian matrix of $f_{\mathsf{A}}(s)$ works out to be:

$$\mathbf{J}_{f_{\mathsf{A}}}(s) = -\kappa_0 \begin{bmatrix} 0 & 0 \\ -h'(p^x) & 1 \end{bmatrix}. \tag{5.7}$$

Fig. 5.3 The energy balance along these paths is positive, showing that the corresponding force fields are not conservative

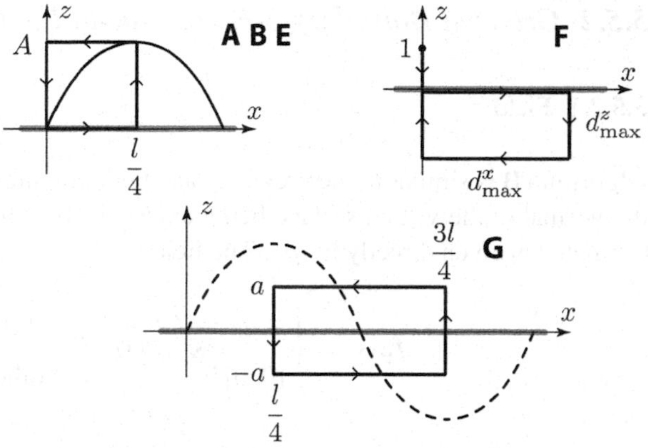

5.5.1.3 Characteristic Number

The norm of $\mathbf{J}_{f_\mathbf{A}}(s)$ is

$$\|\mathbf{J}_{f_\mathbf{A}}(s)\|_2 = \kappa_0\sqrt{1+[h'(p^x)]^2} \tag{5.8}$$

giving

$$q_\mathbf{A} = \kappa_\mathbf{A}^{\max}/\kappa_0 = \sqrt{1+[h'(p^x)]^2} \tag{5.9}$$

that expresses the increase in stiffness incurred by simulating a virtual texture with Algorithm **A**.

For a sinusoidal surface $h(d^x) = A\sin(2\pi p^x/l)$ the maximum value of the norm of the Jacobian matrix is:

$$\kappa_\mathbf{A}^{\max} = \kappa_0\sqrt{1+[2\pi A/l]^2} \tag{5.10}$$

when $p^x = 0 \mod (\pi/l)$. Since typically, $2\pi A/l \gg 1$, the gain is nearly proportional to κ_0, A, and $1/l$. Thus, it has a form similar to that of the Condition 5 of Sect. 5.3.

5.5.1.4 Conservativity

By inspection of (5.7) we can conclude that the curl is not zero, hence, the algorithm does not generate a conservative field.

For example, with $h(p^x) = A\sin(2\pi p^x/l)$, when traversing the closed path $(0,0) \to (l/4,0) \to (l/4,A) \to (0,A) \to (0,0)$, see Fig. 5.3, the virtual environment will generates energy $\Delta E = +1/2\kappa_0 A^2$.

5.5.2 Grooved Boundary—Force Normal to Groove (B)

5.5.2.1 Field

Algorithm **B** is similar to the previous one, but computes a force that is aligned with the normal to the virtual surface being explored [6]. The point where the normal is evaluated is given directly by p^x. The field is

$$f_{\mathsf{B}}(s) = \begin{cases} -\kappa_0 d^z \frac{[-h'(p^x),1]^\top}{\sqrt{1+h'(p^x)^2}} & \text{if } d^z < 0, \\ [0,0]^\top & \text{otherwise.} \end{cases} \tag{5.11}$$

5.5.2.2 Jacobian and Characteristic Number

A manageable closed form could not be found but can be numerically evaluated.

5.5.2.3 Conservativity

When $h(p^x) = A\sin(2\pi p^x/l)$, cycling through the path $(0,0) \to (\frac{l}{4},0) \to (\frac{l}{4},A) \to (0,A) \to (0,0)$ generates

$$\Delta E = \frac{\kappa_0 A^2}{2} + \frac{\kappa_0 L^2}{4\pi^2}\left(1 - \sqrt{\frac{L^2 + 4A^2\pi^2}{L^2}}\right). \tag{5.12}$$

Remark 5.4 Algorithms **A** and **B** do not generate conservative force fields and are known to cause a feel of activity.

5.5.3 Change of Height (C)

Algorithm **C**, described in [17], is the extension of that used in the Sandpaper System to three dimensions [26]. The force field does not follow the normal of the texture; it depends on the position but not on the trajectory of the handle. A related algorithm, presented in [21], computes a force with a tangential component that is proportional to the rate of change of the height of the virtual textured surface with respect to the curvilinear abscissa of the scan path. The algorithm results from the observation that the intensity of an interaction force should relate both to the shape of the surface and to the manner in which it is explored. It was observed that this algorithm can be viewed as a way to eliminate non-working forces [13]. In this paper we refer to the static version of this algorithm [17].

5.5.3.1 Field

Given a height map $h(p^x)$, a potential U_{C} function of the normal deflection can be defined:

$$U_{\mathsf{C}} = \begin{cases} -\kappa_0 [d^z]^2/2 & \text{if } d^z < 0, \\ 0 & \text{otherwise.} \end{cases} \tag{5.13}$$

The gradient of U_{C} gives the force field

$$f_{\mathsf{C}}(s) = \begin{cases} -\kappa_0 [-d^z h'(p^x), d^z]^\top & \text{if } d^z < 0, \\ [0, 0]^\top & \text{otherwise} \end{cases} \tag{5.14}$$

which extends the rendering described in [21] to account for normal deflection.

5.5.3.2 Jacobian

The Jacobian matrix of this field

$$\mathbf{J}_{f_{\mathsf{C}}}(s) = -\kappa_0 \begin{bmatrix} [h'(p^x)]^2 - d^z h''(p^x) & -h'(p^x) \\ -h'(p^x) & 1 \end{bmatrix} \tag{5.15}$$

depends on the second derivative of the height field h. This fact suggests that the texture stiffness grows with the *curvature* of the simulated surface. This actually has a nice physical interpretation since we would expect the interaction forces, and hence the gain of the synthesis, to become very large in the sharp asperities and crevices of an irregular surface.

5.5.3.3 Characteristic Number

It must be numerically evaluated.

5.5.3.4 Conservativity

Given by construction.

Remark 5.5 A key difference between algorithms **B** and **C** is the normalization factor $\sqrt{1 + (h'(p^x)^2)}$ in (5.11). This factor prevents algorithm **B** from generating a conservative force field whereas algorithm **C**'s field is conservative.

5.5.4 Variant 1 Derived from the 'God-Object' Method (D)

Algorithm **D** is a first variant of the 'god-object' method [36], which at all times minimizes the amount of penetration inside a virtual surface.

5.5.4.1 Field

If p_h is the surface point closest to p, given a texture with a generating function $p_h^z = h(p_h^x)$, a potential U_{D} can be defined

$$U_{\mathsf{D}} = \begin{cases} -\kappa_0[(p^x - p_h^x)^2 + (p^z - p_h^z)^2]/2 & \text{inside,} \\ 0 & \text{outside,} \end{cases} \tag{5.16}$$

where (p_h^x, p_h^z) are the coordinates of p_h.

By taking the gradient, we obtain

$$f_{\mathsf{D}}(s) = \begin{cases} -\kappa_0[(p^x - p_h^x), (p^z - p_h^z)]^\top & \text{inside,} \\ 0 & \text{outside.} \end{cases} \tag{5.17}$$

By virtue of being a gradient, the algorithm is conservative.

The authors found an inaccuracy in the two following sections Jacobian and Characteristic Number. Please refer to Appendix 3 at page 95 for a complete description of the passivity properties of algorithm **D**.

5.5.4.2 Jacobian

The Jacobian matrix inside the boundary is (see Appendix 2)

$$\mathbf{J}_{f_{\mathsf{D}}}(s) = -\frac{\kappa_0}{(1 + h'(p_h^x)^2)} \begin{bmatrix} h'(p_h^x)^2 & -h'(p_h^x) \\ -h'(p_h^x) & 1 \end{bmatrix}. \tag{5.18}$$

This matrix is singular. Moving in a direction orthogonal to the minimum penetration vector keeps the force constant as long as the potential is differentiable.

5.5.4.3 Characteristic Number

Notice that $\|\mathbf{J}_{f_{\mathsf{D}}}(s)\|_2 = \kappa_0$, showing no dependency on the boundary function. In that, its differs from algorithms **A** and **C**. The characteristic number is 1, a nice property indeed.

5.5.4.4 Conservativity

Given by construction.

Remark 5.6 Algorithm **D** has the peculiarity that it synthesizes a discontinuous force field, see Fig. 5.2, thus causing undesirable artifacts. For a sinusoidal texture, the discontinuities occur at the maxima of the sinusoidal function. For a general profile, the artifacts arise when the probe is equidistant from two or more different points of the boundary. The corresponding potential is continuous but not differentiable, even when the boundary is smooth, which is not physical unless shocks are acceptable.

5.5.5 Variant 2 Derived from the 'God-Object' Method (**E**)

Algorithm **E** is a second variant of the 'god object' method.

5.5.5.1 Field

Here, the force field is such that the synthesized force is always normal to surface, Fig. 5.2d. The force field is

$$f_{\mathsf{E}}(s) = \begin{cases} [0, -\kappa_0 \sqrt{(p^x - p_h^x)^2 + (p^z - p_h^z)^2}]^\top & \text{inside,} \\ [0, 0]^\top & \text{outside.} \end{cases} \tag{5.19}$$

5.5.5.2 Jacobian

Inside the boundary, the Jacobian matrix is

$$\mathbf{J}_{f_{\mathsf{E}}}(s) = -\frac{\kappa_0 \begin{bmatrix} 0 & 0 \\ (p^x - p_h^x) & (p^z - p_h^z) \end{bmatrix}}{\sqrt{(p^x - p_h^x)^2 + (p^z - p_h^z)^2}}. \tag{5.20}$$

5.5.5.3 Characteristic Number

The norm of the matrix is

$$\|\mathbf{J}_{f_{\mathsf{E}}}(s)\|_2 = \kappa_0 \tag{5.21}$$

and the characteristic number is also 1.

5.5.5.4 Conservativity

Unfortunately, the form of the Jacobian matrix shows that the force field is non-conservative. In fact, on the closed path (Fig. 5.3) $(0, 0) \to (\frac{l}{4}, 0) \to (\frac{l}{4}, A) \to (0, A) \to (0, 0)$ the energy change is on the segment $(\frac{l}{4}, 0) \to (\frac{l}{4}, A)$ only where force and displacement are aligned. Hence $\Delta E > 0$.

Remark 5.7 This second variant of the 'god-object' method, algorithm **E**, while being non-conservative, does not suffer from the discontinuity problem of algorithm **D**.

5.5.6 Flat Wall with Modulated Lateral Friction (F)

Texture rendering with dry friction is not new, see [32]. Here we introduce formulation based on a time-free friction model (5.22). Instead of adding several contact force components, the tangential friction force is directly modulated by the height field of the texture.

5.5.6.1 Field

A tangential friction force component along x is modulated by a function of the net tangential displacement. The virtual friction is computed using the technique described in [19]. In the formula of Table 5.1, $\kappa(.)$ has the dimension of a stiffness. Specifically, $\kappa(x) = \mu\kappa_0[1 - h(x)]$, where μ is Amontons' coefficient of friction, and the quantity in bracket is the modulation function. The lateral force component is combined with the normal response of an ordinary wall. This results in a synthesis that is independently tunable in the two directions, i.e. a "DC" nominal normal component and a varying, oscillatory lateral component. The force field is

$$f_{\mathsf{F}}(s) = \begin{cases} -\kappa_0[\mu[1 - h(p^x)]\frac{d^x}{d^x_{\max}} p^z, p^z]^\top, & p^z < 0, \\ [0, 0]^\top & \text{otherwise}, \end{cases} \tag{5.22}$$

where $d^x_{\max}$ is the maximum pre-sliding tangential deflection [19]. For this algorithm we assume additionally that $0 \le h(p^x) \le 1$.

5.5.6.2 Jacobian

The Jacobian matrix is:

$$\mathbf{J}_{f_{\mathsf{F}}}(s) = -\kappa_0 \begin{bmatrix} \mu\frac{p^z}{d^x_{\max}}(\frac{\mathrm{d}d^x}{\mathrm{d}p^x} - h(p^x)\frac{\mathrm{d}d^x}{\mathrm{d}p^x} - h'(p^x)) & \mu[1 - h(p^x)]\frac{d^x}{d^x_{\max}} \\ 0 & 1 \end{bmatrix}. \tag{5.23}$$

This algorithm has memory in the term $\mathrm{d}d^x/\mathrm{d}p^x$. The worst case for the condition number of the Jacobian needs to be investigated. According to [19]:

$$\text{stick phase:} \quad \begin{cases} d^x < d^x_{\max} \\ \frac{\mathrm{d}d^x}{\mathrm{d}p^x} = 1 \end{cases}, \qquad \text{slip phase:} \quad \begin{cases} d^x = d^x_{\max} \\ \frac{\mathrm{d}d^x}{\mathrm{d}p^x} = 0 \end{cases}. \tag{5.24}$$

For the example of $h(p^x) = A\sin(2\pi p^x/l)$ we can maximize each entry of the Jacobian matrix.

$$\mathbf{J}_{f_{\mathsf{F}:\mathrm{slip}}}(s) = -\kappa_0 \begin{bmatrix} 2\pi\mu p^z A/l & 2\mu \\ 0 & 1 \end{bmatrix} \tag{5.25}$$

and

$$\mathbf{J}_{f_{\mathsf{F:stick}}}(s) = -\kappa_0 \begin{bmatrix} 2\mu p^z(1/d^x_{\max} + \pi A/l) & 2\mu \\ 0 & 1 \end{bmatrix} \tag{5.26}$$

are the maximum values of the Jacobian matrix during the slip and stick phases respectively.

5.5.6.3 Characteristic Number

The symbolic expression for the norm of $\mathbf{J}_{f_{\mathsf{F:stick}}}$ is cumbersome but the Appendix 1 can be consulted for the expression of $q_{\mathsf{F}} = \|\mathbf{J}_{f_{\mathsf{F}}}\|_2/\kappa_0$.

Notice that in Eq. (5.26) the gain of the lateral force component depends on p^z. Because of the user input, p^z can grow unbounded, giving in principle an arbitrarily large gain. To address this problem, p^z must be clamped to a maximum $d^z_{\max}$. We then have a three-way trade off between A, l and $d^z_{\max}$. Note that realistic values for $d^x_{\max}$ as well as for $d^z_{\max}$ may range from 10^{-2} m (biological tissues) to 10^{-6} m (metals).

5.5.6.4 Conservativity

The Jacobian matrix (5.23) shows that the algorithm does not in general produce a conservative field. The sliding friction is guaranteed to be dissipative but energy can be generated during the stick phase.

On the path $(0, 1) \to (0, -a) \to (d^x_{\max}, -a) \to (d^x_{\max}, -d^z_{\max}) \to (0, -d^z_{\max}) \to (0, 1)$, (Fig. 5.3), and for $A = 0$, the energy gain is

$$\Delta E = +\frac{\kappa_0 \mu}{2} d^z_{\max} d^x_{\max}. \tag{5.27}$$

The algorithm is generative only in the stick phase which is at most $2d^x_{\max}$ wide. This confined energy gain does not create artifacts in the following experiments. However, this gain might be a problem when rendering highly compliant deformable bodies ($d^x_{\max} \approx 1$ cm) because the non-conservative energy is proportional to $d^x_{\max}$.

5.5.7 *Force Shading (G)*

Lastly, the 'force-shading' algorithm G, which synthesizes a force normal to the texture, was implemented [27].

Table 5.2 Summary of Algorithms Properties

	Char. number	Conservativity	Continuity
A	$\sqrt{1+[2\pi A/l]^2}$	No	If h continuous
B	No closed-form	No	If h' continuous
C	No closed-form	Yes	If h' continuous
D	1	Yes	No. Even if h continuous
E	1	No	If h continuous
F	Appendix 1	Almost	If h continuous
G	No closed-form	No	If h' continuous

5.5.7.1 Field

The field is

$$f_{\mathsf{G}}(s) = \begin{cases} -\kappa_0 p^z \frac{[-h'(p^x),1]^\top}{\sqrt{1+(h'(p^x))^2}} \boldsymbol{n} & \text{if } p^z < 0, \\ [0,0]^\top & \text{otherwise.} \end{cases} \tag{5.28}$$

5.5.7.2 Jacobian

The Jacobian matrix does not easily afford a closed-form solution.

5.5.7.3 Characteristic Number

While an expression could not be found for the algorithm's characteristic number, it was possible to tune it empirically.

5.5.7.4 Conservativity

On the path $(l/4, a) \to (l/4, -a) \to (3l/4, -a) \to (3l/4, a) \to (l/4, a)$ for example, (Fig. 5.3), the algorithm is non-conservative because the energy gain is

$$\Delta E = -1/2\kappa_0 a^2 + \Delta E_2 + 1/2\kappa_0 a^2 + 0 = \Delta E_2 > 0 \tag{5.29}$$

since force and the displacement are aligned only on the segment $(l/4, -a) \to (3l/4, -a)$.

5.5.8 Summary

Table 5.2 summarizes the discussion of the previous sections.

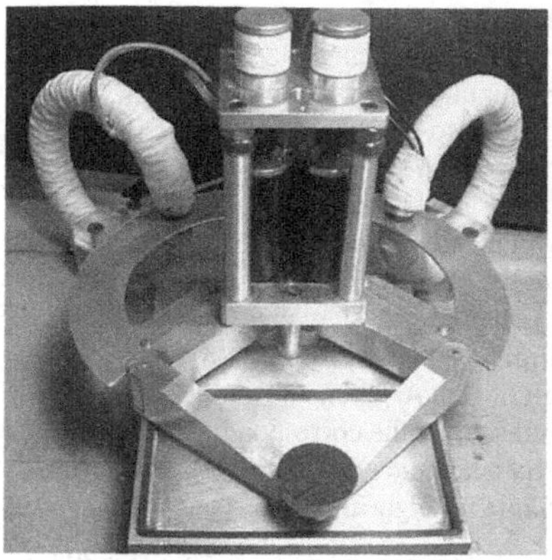

Fig. 5.4 The 2D device used in the experiments. It has two direct-drive motors and two programmable eddy-current brakes. An elastic band (not shown) is attached to the handle during the passivity experiments

5.6 Experimental Validation

In a first step, we use an elastic rubber band to verify that, indeed, the mere fact of adding texture to an otherwise smooth virtual surface does increase its apparent stiffness and may lead to a loss of passivity.

In a second step, an operator actively explores textured surfaces. The energy supplied to the haptic device is recorded. The reason behind the second set of tests is that lack of conservativity of the underlying force field, can only be revealed through active scanning of the simulated surface. It can be verified that, as predicted, different algorithms have radically different energetic behaviors when tested under similar conditions.

To perform these tests under the best conditions possible, we use a two degrees-of-freedom device, which was rigorously characterized [5]. From these tests, we established that its spatial resolution measured by external backdriving is better than 15 μm. The sampling period is $\mathcal{T} = 10^{-4}$ s and the highest texture spatial period is $l = 1$ mm. Since the scanning velocity is smaller than 0.2 m^{-1}, the number of samples α and β is always larger than 50 and the maximum frequency of 200 Hz is well within the 400 Hz-wide 3 dB flatband of the device [5]. For the experiments, we use the variant in [16], which enables us to program high quality viscous damping in the machine, see Fig. 5.4.

5.6.1 Passivity Experiments

A virtual wall, algorithm **W**, was set up with the force field

$$f_{\mathbf{W}}(s) = \begin{cases} [0, -\kappa_0 p^z]^\top & \text{if } p^z < 0, \\ [0, 0]^\top & \text{otherwise.} \end{cases} \tag{5.30}$$

The intrinsic physical dissipation in the device is an imperceptible amount of friction in the joints of device, internal friction in the rubber band, and air drag

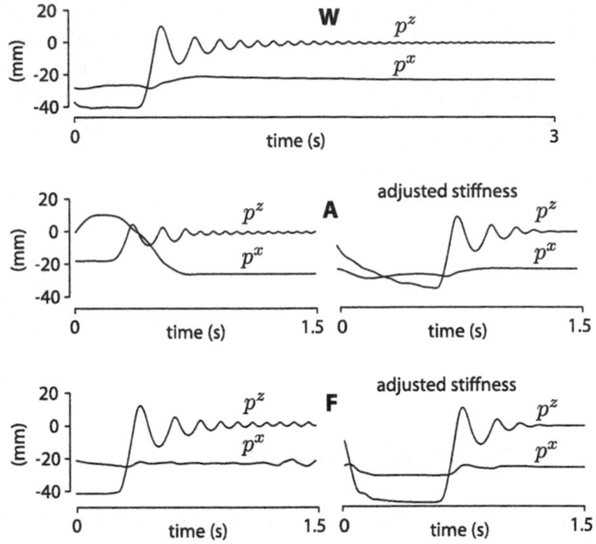

Fig. 5.5 Plots were obtained by thrusting the manipulandum with an elastic band onto virtual walls. For the wall, **W**, algorithm **A** (*left*), and algorithm **F** (*left*), when $\kappa_0 = 2000\ \mathrm{N\,m^{-1}}$. The wall is passive but once textured, the system enters a limit cycle. Behavior of algorithm **A**, when κ_0 is reduced by the corresponding characteristic number (*right*). Same for algorithm **F**

around the linkages. We also add viscous damping when the handle is inside the virtual object, 4.7 mN m s in each joint. Next, the manipulandum is thrust into the wall such that a lack of passivity is indicated by the onset of spontaneous activity. Referring to Fig. 5.5, stiffness is tuned so that the smooth virtual wall is passive. When exactly the same wall is textured with algorithm **A** or **F**, stable limit cycles occur. The conditions are $\kappa_0 = 2000\ \mathrm{N\,m^{-1}}$, $l = 1$ mm, $A = 0.8$ mm, $\mu = 0.8$, $d^x_{\max} = d^z_{\max} = 1$ mm. We then apply the theory of Sect. 5.4 to show that if we account for the characteristic numbers of algorithms, $q_{\mathsf{F}} = 8.3$ and $q_{\mathsf{A}} = 5.1$, we can ensure passivity.

5.6.2 Conservativity Experiments

A subject explores a virtual wall textured with selected algorithms discussed earlier and their energetic behavior is recorded. In principle, to test conservativity, the system should follow closed paths. Admittedly, it is rather hard to ask a subject to do this accurately. In practice, we can replace cycling through a closed trajectory by oscillations around a nominal value and monitor the *average* work made during many cycles. Leaving everything else unchanged, we can compare algorithms by computing $\sum_0^N f_x \Delta p^x$ and $\sum_0^N f_z \Delta p^z$ where Δp^x and Δp^y are the incremental displacements of the position observed during one sample period, that is, every 100 μs in the present experiments. The quantities f_z and f_x are the force commands sent to the device.

A textured wall is set up with parameters $\kappa_0 = 1000\ \mathrm{N\,m^{-1}}$, $l = 2$ mm, $A = 0.8$ mm, $\mu = 0.8$, $d^x_{\max} = d^z_{\max} = 1$ mm. The analog dampers are energized when the handle is inside the virtual body to 4.0 mN m s. The passivity margin of this

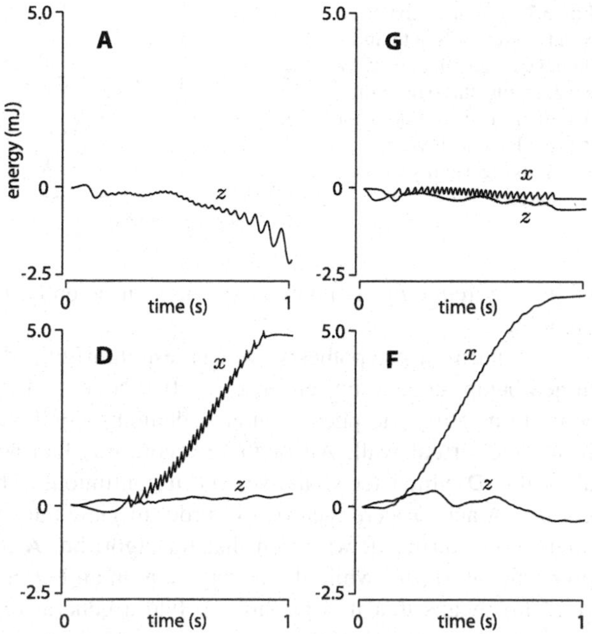

Fig. 5.6 Plots obtained by computing the energy supplied to the device dragging the handle along a textured wall. The user moved the handle while attempting to keep the penetration as constant as possible

experimental condition is larger than the one shown for the passivity tests. Following the usual convention, a decreasing energy indicates that the energy is supplied by the virtual environment. Figure 5.6 shows the energy balance during a single stroke along the texture that lasts approximately 1 second.

With algorithm **A** there is a noticeable energy gain in the z direction indicative of a generative behavior during a single continuous lateral motion. There is no energy associated with the lateral direction. By comparison, the 'force-shading' algorithm **G** is marginally passive in the normal as well as in the lateral directions although it is not strictly conservative. Algorithm **D**, employing the 'god-object' approach, is marginally passive in the normal direction z and is clearly passive in the lateral direction x for an equivalent textural vibration. This was despite the fact that algorithm **D** synthesizes a discontinuous force field that causes the hills to feel 'clipped' and distorted. Algorithm **F** is marginally passive in the normal direction z and is passive in the lateral direction x as a result of the dissipative nature of the friction model.

5.6.3 Surface Activity

Choi and Tan referred to low frequency vibrations experienced by users when exploring surfaces synthesized by algorithms **A** and **B** [6]. This artifact is likely to be associated to the fact that the force fields generated by these algorithms are not conservative. In our experience, this sensation arises even if the virtual surface is made

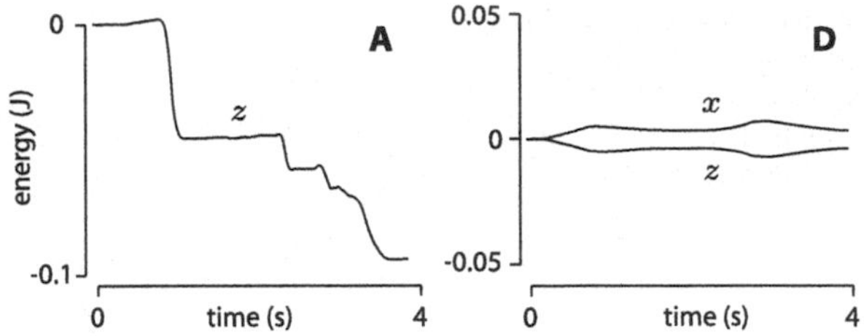

Fig. 5.7 The microdynamics of interaction with a texture boundary were magnified by exaggerating the size of the feature. Here, a 20 mm bump was explored and two synthesis algorithms were compared

passive by reducing its nominal stiffness sufficiently, or even by adding physical damping.

To confirm this hypothesis, we can test algorithms **A** , **B**, **E**, and **D** with an 'enlarged bump' created by setting $\kappa_0 = 1000\ \mathrm{N\,m^{-1}}$, $A = 20$ mm and $l = 20$ mm in order to magnify the phenomenon; a damping coefficient of 4.0 mN m s is added inside the virtual wall. All of these algorithms feel active with the exception of algorithm **D** which feels passive, yet discontinuous. The results of employing algorithm **A** and **D** were recorded in order to view the 'magnified' dynamics of the interaction. Earlier, it was seen that for algorithm **A** the characteristic number is proportional to A/l while the energy gain on an horizontal path is proportional to A^2. This means that it is possible to find a generative texture algorithm that has passivity characteristics. The results are in Fig. 5.7.

Algorithm **A**, because it 'pushes' the handle in a direction that differs from its movements, makes it difficult for the user to interact with the virtual object. Contact is often lost as a result of the sharp energy increase created at the contact with the surface. The energy plot, Fig. 5.7, shows these energy jumps. Algorithm **D**, however, maintains an overall energy balance. From these tests, we surmise that algorithms which yield non-conservative fields feel 'alive' although, locally, the complete system may be passive, recall Sect. 5.4.3. In particular, algorithm **A** is passive when the probes moves perpendicularly to the surface as shown in the experiment with the elastic band.

5.7 Conclusion

The properties of common texture synthesis algorithms were analyzed in terms of their effective stiffness, that is, of the incremental stiffness they cause. Texturing a surface generally decreases the passivity margin of a given virtual mechanical environment. We found that any specific algorithm has a characteristic number that expresses this gain succinctly. In some cases this number is constant or even equal to 1. In others cases, it depends on the texture parameters and is most frequently of the form $\propto A/l$, where A is the height of the texture and l its spatial period.

It is confirmed that absence of surface activity and passivity are related but distinct notions. Surface activity is related to unphysical artifacts generated by algorithms synthesizing non-conservative force fields. Passivity is related to the system dynamics and can always be ensured by reducing the closed loop gain, by increasing the sample rate, or by adding dissipative elements. Conservativity, on the other hand, is an intrinsic property of algorithmic synthesis which may or may not be ensured, independently of any particular hardware or control technique.

We demonstrated that some widely used texturing algorithms do create non-conservative force fields for typical exploration paths, linking energy gain to artifacts which are not intended by the application programmer. We showed that continuous and conservative texture force fields can be created using a change-of-height approach. The 'god-object' approach was found to be nominally conservative. It leaves the effective stiffness equal to the original stiffness but produces undesirable discontinuities. A variant that creates a force field that is always normal to the original surface cured this problem but caused the algorithm produce non-conservative force fields and hence active surfaces. We introduced a new synthesis algorithm which creates textures by modulating the lateral friction force when scanning a surface. It provides parameters that can be tuned to deliver realistic sensations and a simple formulation for its characteristic number. This algorithm is not particularly advantageous in terms of its effective stiffness but the dissipative nature of the friction field makes the texture feel realistic and passive.

Finally, the perceptual qualities of the various algorithms remain to be investigated with regard to new haptic device interface concepts such as that in [16, 33].

Acknowledgements The authors would like to thank Mohsen Mahvash and Andrew H.C. Gosline for insightful comments on earlier drafts of this paper. This research was supported in part by the Institute for Robotics and Intelligent Systems, and NSERC, the Natural Sciences and Engineering Research Council of Canada.

Appendix 1: Characteristic Number of Algorithm F

An upper bound for q_{F} $h(p^x) = A\sin(2\pi p^x/l)$ is

$$
\begin{aligned}
q_{\mathsf{F}} \le {} & \left\| \begin{matrix} 2\mu p^z(1/d^x_{\max} + \pi A/l) & 2\mu \\ 0 & 1 \end{matrix} \right\|_2 \\
\le {} & \frac{\sqrt{2}}{2d^x_{\max}} \Big[4(\mu p^z)^2 + 8(\mu p^z)^2 \pi A d^x_{\max}/l \\
& + 4(\mu p^z \pi A d^x_{\max}/l)^2 + 4(\mu d^x_{\max})^2 + {d^x_{\max}}^2 \\
& + \big(16(\mu^4 p^z)^4 + 32\mu^2(\mu^2 d^x_{\max} p^z)^2 \\
& + 64(\mu p^z)^4 \pi A d^x_{\max}/l + {d^x_{\max}}^4 + 96(\mu p^z)^4(\pi A d^x_{\max}/l)^2 \\
& + 64(\mu p^z)^4(\pi A d^x_{\max}/l)^3 + 64\mu^4 {d^x_{\max}}^3 {p^z}^2 \pi A/l \\
& - 16{d^x_{\max}}^3(\mu p^z)^2 \pi A/l + 16(\mu p^z \pi A d^x_{\max}/l)^4 \\
& + 32(\mu^4 d^x_{\max})^4(p^z \pi A/l)^2 - 8{d^x_{\max}}^4(\mu p^z \pi A/l)^2 \\
& + 16(\mu d^x_{\max})^4 + 8\mu^2 {d^x_{\max}}^4 - 8(\mu p^z d^x_{\max})^2\big)^{1/2}\Big]^{1/2}.
\end{aligned}
\tag{5.31}
$$

Appendix 2: Jacobian Matrix of Algorithm D

For the god-object method, the assumptions are:

- the boundary curve $h(p^x)$ is smooth and differentiable.
- there is just one point (p_h^x, p_h^z) on h that minimizes the distance between the (p^x, p^z) and the boundary.
- probe is 'inside' the texture.

We know that:

$$(p^z - p_h^z) = -(p^x - p_h^x)/h'(p_h^x) \tag{5.32}$$

while the linearization around (p^x, p^z) gives:

$$\frac{\partial p_h^z}{\partial p^x} = h'(p_h^x)\frac{\partial p_h^x}{\partial p^x}, \tag{5.33}$$

$$\frac{\partial p_h^x}{\partial p^z} = \frac{\partial p_h^z}{\partial p^z}/h'(p_h^x). \tag{5.34}$$

The energy function can be described by

$$E_{\mathsf{D}} = -\kappa_0[(p^x - p_h^x)^2 + (p^z - p_h^z)^2]/2. \tag{5.35}$$

Differentiating (5.35) gives

$$f_{\mathsf{D}}(s)_x = \frac{\partial E_{\mathsf{D}}}{\partial p^x} \tag{5.36}$$

$$= -\kappa_0\left((p^x - p_h^x)\left(1 - \frac{\partial p_h^x}{\partial p^x}\right) - (p^z - p_h^z)\frac{\partial p_h^z}{\partial p^x}\right)$$

$$= -\kappa_0(p^x - p_h^x) + \kappa_0\left(\frac{\partial p_h^x}{\partial p^x} - \frac{\partial p_h^z}{\partial p^x}/h'(p_h^x)\right) \tag{5.37}$$

$$= -\kappa_0(p^x - p_h^x). \tag{5.38}$$

where we used Eq. (5.32), (5.33), and (5.34). The derivation of the force is complete by defining $f_{\mathsf{D}}(s)_x = 0$ if $h'(p_h^x) = 0$.

The same reasoning can be used to derive

$$f_{\mathsf{D}}(s)_z = -\kappa_0(p^z - p_h^z) \tag{5.39}$$

and $f_{\mathsf{D}}(s)_z = 0$ if $1/h'(p_h^x) = 0$. The Jacobian can be easily computed:

$$\mathbf{J}_{f_{\mathsf{D}}}(s) = -\kappa_0 \begin{bmatrix} 1 - \frac{\partial p_h^x}{\partial p^x} & -\frac{\partial p_h^x}{\partial p^z} \\ -\frac{\partial p_h^z}{\partial p^x} & 1 - \frac{\partial p_h^z}{\partial p^z} \end{bmatrix} \tag{5.40}$$

hence

$$\mathbf{J}_{f_\mathsf{D}}(s) = -\kappa_0 \begin{bmatrix} 1 - \frac{\partial p_h^x}{\partial p^x} & -\frac{\partial p_h^z}{\partial p^z}/h'(p_h^x) \\ -h'(p_h^x)\frac{\partial p_h^x}{\partial p^x} & 1 - \frac{\partial p_h^z}{\partial p^z} \end{bmatrix}. \tag{5.41}$$

Knowing that:

$$\frac{\partial p_h^x}{\partial p^x} = \cos(\mathrm{atan}(h'(p_h^x)))^2 = 1/(1 + h'(p_h^x)^2) \tag{5.42}$$

and

$$\frac{\partial p_h^z}{\partial p^z} = \sin(\mathrm{atan}(h'(p_h^x)))^2 = h'(p_h^x)^2/(1 + h'(p_h^x)^2) \tag{5.43}$$

we can write

$$\mathbf{J}_{f_\mathsf{D}}(s) = -\frac{\kappa_0}{1 + h'(p_h^x)^2} \begin{bmatrix} h'(p_h^x)^2 & -h'(p_h^x) \\ -h'(p_h^x) & 1 \end{bmatrix}. \tag{5.44}$$

Appendix 3: Erratum to "On the Synthesis of Haptic Textures"[1]

In Sect. IV.D.3 of reference [4] (5.5.4 of this book) it is stated that the characteristic number of the first variant of the 'god-object' method applied to texture synthesis is 1. However, the assumptions used to simplify the Jacobian matrix neglect the effect of curvature of the height function, hence this statement is not always true. The 'god-object' method generates a force field based on minimizing the distance between the interaction point, $p = (p^x, p^z)$, and a point $p_h = (p_h^x, p_h^z)$ on the texture, $z = h(x)$. In other words, there exists a function $u(x, y)$ such that $p_h^u, p_h^{h(u)}$ minimizes $\|p - p_h\|$ and such that the vector $p_h - p$ is normal to the curve h at point u. In a reference frame at located at u (tangent and normal to the curve, coordinates ξ, ν, see Fig. 5.1), the complete expression of the Jacobian matrix of the force field is (the original expression was given in global coordinates)

$$\kappa_0 \begin{pmatrix} h''(0)\nu/(1 - h''(0)\nu) & 0 \\ 0 & -1 \end{pmatrix}. \tag{5.45}$$

Upon evaluation of the norm of this matrix, it can be found that the characteristic number is indeed 1 except when the interaction point approaches the center of

[1]Reprinted from Gianni Campion and Vincent Hayward, Erratum to "On the Synthesis of Haptic Textures." *IEEE Transactions on Robotics*, Volume 25, Issue 2, 475, 2008.

curvature of the curve describing the height function. Specifically, it is

$$1 \quad \text{if } h''(0) \geq 0, \tag{5.46}$$

$$1 \quad \text{if } h''(0) < 0 \wedge \nu \geq 1/(2h''(0)), \tag{5.47}$$

$$|h''(0)\nu/(1 - h''(0)\nu)| \quad \text{if } h''(0) < 0 \wedge \nu < 1/(2h''(0)) \tag{5.48}$$

which has a singularity for $\nu = 1/h''(0)$ and tends to 1 as $\nu \to \infty$. In practice not all centers of curvature represent a problem. For a sine wave $A\sin(2\pi p^x/l)$ there is one singular point per period at $(l, 1 - (l/2\pi)^2)$. Notice that as the spatial frequency increases the singular points are closer to the surface, becoming less likely to be encountered. Moreover, the increase of stiffness happens only in the convex regions where limit cycles are less likely to occur.

References

1. Abbott, J.J., Okamura, A.M.: Effects of position quantization and sampling rate on virtual wall passivity. IEEE Trans. Robot. **21**(5), 952–964 (2005)
2. Adams, R.J., Hannaford, B.: Stable haptic interaction with virtual environments. IEEE Trans. Robot. Autom. **15**(3), 465–474 (1999)
3. Campion, G., Hayward, V.: Fundamental limits in the rendering of virtual haptic textures. In: Proceedings of the First Joint Eurohaptics Conference and Symposium on Haptic Interfaces for Virtual Environments and Teleoperator Systems, WHC'05, pp. 263–270 (2005)
4. Campion, G., Hayward, V.: On the synthesis of haptic textures. IEEE Trans. Robot. **24**(3), 527–536 (2008)
5. Campion, G., Wang, Q., Hayward, V.: The Pantograph Mk-II: A haptic instrument. In: Proceedings of the IEEE/RSJ International Conference on Intelligent Robots and Systems, IROS'05, pp. 723–728 (2005)
6. Choi, S., Tan, H.Z.: Perceived instability of virtual haptic texture. I. Experimental studies. Presence **13**(4), 395–415 (2004)
7. Choi, S., Tan, H.Z.: Perceived instability of virtual haptic texture. II. Effect of collision-detection algorithm. Presence **14**(4), 463–481 (2005)
8. Colgate, J.E., Schenkel, G.: Passivity of a class of sampled-data systems: Application to haptic interfaces. In: Proceedings of the American Control Conference, pp. 3236–3240 (1994)
9. Colgate, J.E., Grafing, P.E., Stanley, M.C., Schenkel, G.: Implementation of stiff virtual walls in force-reflecting interfaces. In: Virtual Reality Annual International Symposium, pp. 202–208 (1993)
10. Costa, M.A., Cutkosky, M.R.: Roughness perception of haptically displayed fractal surfaces. In: Proceedings ASME IMECE Symposium on Haptic Interfaces for Virtual Environments and Teleoperator Systems, vol. 69-2, pp. 1073–1079 (2000)
11. Crossan, A., Williamson, J., Murray-Smith, R.: Haptic granular synthesis: Targeting, visualisation and texturing. In: Proceedings of the International Symposium on Non-visual & Multimodal Visualization, pp. 527–532. IEEE Press, New York (2004)
12. Diolaiti, N., Niemeyer, G., Barbagli, F., Salisbury, J.K.: Stability of haptic rendering: Discretization, quantization, time delay, and coulomb effects. IEEE Trans. Robot. **22**(2), 256–268 (2006)
13. Frisoli, A.: Personal communication (2004)
14. Fritz, J.P., Barner, K.E.: Stochastic models for haptic textures. In: Stein, M.R. (ed.) Telemanipulator and Telepresence Technologies III. Proc. SPIE, vol. 2901, pp. 34–44 (1996)
15. Goldstein, H.: Classical Mechanics. Addison-Wesley, Reading (1950)

16. Gosline, A.H., Campion, G., Hayward, V.: On the use of eddy current brakes as tunable, fast turn-on viscous dampers for haptic rendering. In: Proceedings of Eurohaptics, pp. 229–234 (2006)
17. Hardwick, A., Furner, S., Rush, J.: Tactile display of virtual reality from the world wide web—a potential access method for blind people. Displays **18**, 153–161 (1998)
18. Hayward, V.: Haptic synthesis. In: Proceedings of the 8th International IFAC Symposium on Robot Control, SYROCO 2006, pp. 19–24 (2006)
19. Hayward, V., Armstrong, B.: A new computational model of friction applied to haptic rendering. In: Corke, P., Trevelyan, J. (eds.) Experimental Robotics VI. Lecture Notes in Control and Information Sciences, vol. 250, pp. 403–412 (2000)
20. Hayward, V., Astley, O.R.: Performance measures for haptic interfaces. In: Giralt, G., Hirzinger, G. (eds.) Robotics Research: The 7th International Symposium, pp. 195–207. Springer, Heidelberg (1996)
21. Hayward, V., Yi, D.: Change of height: An approach to the haptic display of shape and texture without surface normal. In: Siciliano, B., Dario, P. (eds.) Experimental Robotics VIII. Springer Tracts in Advanced Robotics, pp. 570–579. Springer, Heidelberg (2003)
22. Hill, D.J., Moylan, P.J.: Dissipative dynamical systems: Basic input-output and state properties. J. Franklin Inst. **309**(5), 327–357 (1980)
23. Mahvash, M., Hayward, V.: High fidelity haptic synthesis of contact with deformable bodies. IEEE Comput. Graph. Appl. **24**(2), 48–55 (2004)
24. Mahvash, M., Hayward, V.: High fidelity passive force reflecting virtual environments. IEEE Trans. Robot. **21**(1), 38–46 (2005)
25. Melder, N., Harwin, W.S.: Force shading and bump mapping using the friction cone algorithm. In: Proceedings of the First Joint Eurohaptics Conference and Symposium on Haptic Interfaces for Virtual Environments and Teleoperator Systems, WHC'05, pp. 573–575 (2005)
26. Minsky, M., Lederman, S.J.: Simulated haptic textures: Roughness. In: Proceedings of the ASME IMECE Symposium on Haptic Interfaces for Virtual Environments and Teleoperator Systems, vol. DSC-Vol. 58, pp. 421–426 (1996)
27. Morgenbesser, H.B., Srinivasan, M.A.: Force shading for haptic shape perception. In: Proceedings of the Fifth Symposium on Haptic Interfaces for Virtual Environments and Teleoperators, ASME Dynamic Systems and Control Division, vol. DSC 58, pp. 407–412 (1996)
28. Otaduy, M.A., Jain, N., Sud, A., Lin, M.C.: Haptic display of interaction between textured models. In: Proceedings of IEEE Visualization, pp. 297–304 (2004)
29. Otaduy, M.A., Lin, M.C.: A perceptually-inspired force model for haptic texture rendering. In: Proceedings of the 1st Symposium on Applied Perception in Graphics and Visualization, pp. 123–126. ACM Press, New York (2004)
30. Salisbury, J.K., Conti, F., Barbagli, F.: Haptic rendering: Introductory concepts. IEEE Comput. Graph. Appl. **24**(2), 24–32 (2004)
31. Salisbury, K.J., Brock, D., Massie, T., Swarup, N., Zilles, C.: Haptic rendering: Programming touch interaction with virtual objects. In: Proceedings Symposium on Interactive 3D Graphics, pp. 123–130. ACM Press, New York (1995)
32. Siira, J., Pai, D.K.: Haptic textures—a stochastic approach. In: Proceedings of IEEE International Conference on Robotics and Automation, pp. 557–562 (1996)
33. Wall, S.A., Harwin, W.S.: Effects of physical bandwidth on perception of virtual gratings. In: Proceedings of the Symposium on Haptic Interfaces for Virtual Environments and Teleoperators, ASME Dynamic Systems and Control Division, pp. 1033–1039 (2000)
34. Weisenberger, J.M., Kreier, M.J., Rinker, M.A.: Judging the orientation of sinusoidal and square-wave virtual gratings presented via 2-DOF and 3-DOF haptic interfaces. Haptics-e **1**(4) (2000), online
35. Zhou, K., Doyle, J.C.: Essentials of Robust Control. Prentice Hall, New York (1997)
36. Zilles, C.B., Salisbury, J.K.: A constraint-based god object method for haptic display. In: Proceedings of the IEEE/RSJ International Conference on Intelligent Robots and Systems, IROS'95, vol. 3, pp. 146–151 (1995)

Chapter 6
Passive Realization of Nonlinear Virtual Environments

Abstract Passivity theory applied to haptics is for the most part restricted to the case of linear, one-dimensional virtual environments. Here, the application of passivity theory is extended to account for nonlinear, multidimensional, non-conservative virtual environments with quantization and delay. To analyze these more general cases, the notion of *passive realization*, stricter than passivity, is introduced. This notion is exemplified with multidimensional linear virtual environments and with experiments involving nonlinear environments.

6.1 Introduction

The purpose of this chapter is to introduce the notion of *passively realized virtual environments*. Our motivation arises from the fact that when active organisms, which we are, interact with real environments, this interaction can in general be described by force fields, but real life fields are in general very complicated. They are typically nonlinear, distributed, time-varying and non-conservative. The purpose of virtual environments is to reproduce certain desired aspects of these force fields that can effectively be realized by computer-controlled hardware.

Among all existing approaches, the use of force-feedback devices is by far the most developed. Our results are aimed at those force-feedback devices which operate through the specification of impedance. Using similar methods, however, a counterpart theory could be developed for those operating through specification of admittance.

A virtual environment is said to be 'passively realized' if the haptic device that is generating the forces never returns more energy than what the virtual environment prescribes. In other words, not all the virtual energy that the virtual environment generates is converted in mechanical energy at the handle of the haptic device, but part of it is dissipated by the haptic device.

Our main result, which builds on Theorem 1 of reference [11], is a general expression that can be specialized to specific cases, including those environments that can be locally linearized, which are conservative but realized with delay in the loop. Under proper assumptions and for one single dimension, our result coincides with those found in [1, 6] and hence generalizes these earlier results to several dimensions.

G. Campion, *The Synthesis of Three Dimensional Haptic Textures: Geometry, Control, and Psychophysics*, Springer Series on Touch and Haptic Systems,
DOI 10.1007/978-0-85729-576-7_6, © Springer-Verlag London Limited 2011

Previous approaches did not distinguish between conservative and non-conservative virtual environments [4]. The implications of such a distinction are addressed in the present chapter and exemplified experimentally.

The most important point in this chapter is a further understanding of the concept of characteristic number and its role in describing haptic textures; in particular, this chapter validates the previous understanding that the characteristic number is not influenced by the energetic nature of the texturing algorithms but can be used to compare conservative and non conservative algorithms alike.

6.2 Related Work

The problem of synthesizing virtual mechanical environments that remain passive once realized by a haptic force-feedback system has been widely investigated, particularly for the interaction with a virtual wall along one single direction.

A necessary and sufficient condition for passivity relating the stiffness of a virtual wall to the physical damping in a haptic device was derived in [5]. This analysis has been recently extended to account for time sampling and measurement quantization [1, 6], and the stability conditions for damped virtual walls have been established [7, 8].

In [12], the nonlinearities of the virtual environments were analyzed considering position, velocity, and acceleration dependent forces. Other approaches include those described in reference [2], and more recently in [13]. These approaches ensure passivity by modifying the transfer of force from the virtual environment to the haptic device by means of filters that limit the impedance of the virtual environment or that dissipate excess interaction energy by injecting damping by feedback.

These approaches work in general against the objective of designing high-fidelity virtual environments that synthesize, unmodified, a prescribed interaction field. Alternative methods that do not rely on modifying the prescribed field include exact port-Hamiltonian approaches that operate by integrating the discrete system at each step to obtain matching conditions [14]. Also, an output-limiting approach that can maintain fidelity, while at the same time preserve passivity was recently introduced [9].

There is however in general little work on the extension to the case of nonlinear and multidimensional virtual environments which are needed to create realistic simulations.

6.3 Passively Realized Virtual Environments

6.3.1 Conservativity and Passivity

Definition 6.1 (Passivity) The haptic interaction between a user and a virtual environment is passive if there is a constant β such that

$$\int_0^t \boldsymbol{f}(\tau)^\top \boldsymbol{v}(\tau)\mathrm{d}\tau \geq \beta, \quad \forall t, \tag{6.1}$$

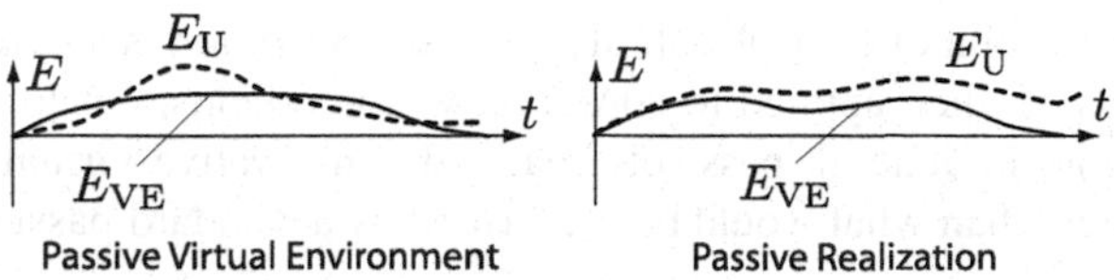

Fig. 6.1 Energy behavior of a passive interaction (*left*) and of a passively realized VE (*right*), where E_{VE} represents the change in virtual energy and E_U represents the energy exchange with the user. Energy with positive sign is dissipated by the haptic device

and for every force field $\boldsymbol{f}(t)$ and allowable trajectory $\boldsymbol{v}(t)$ (adapted from [10]). Please notice that in this formalism the energy with positive sign is the energy dissipated by the haptic device.

A great deal of research on haptic simulation is aimed at specifying virtual force fields, $\boldsymbol{f}(\cdot)$, that can be passively synthesized using a haptic device having dissipative characteristics. By physics, all open-loop haptic devices are dissipative, no matter how small are the losses that they exhibit. However, without consideration of the dissipation due to the device electromechanics, non-conservative force fields may not give rise to passive interaction; in fact, if there exists a closed trajectory $\boldsymbol{v}^*(t)$ such that $\int_0^t \boldsymbol{f}(\tau)^\top \boldsymbol{v}^*(\tau)\mathrm{d}\tau < 0$, then the force field $\boldsymbol{f}(\cdot)$ may not be experienced to be passive [4].

As a result, when complex interactions are desired, standard passivity-based approaches cannot be applied to analyze the haptic interaction. It is therefore necessary to define passivity in a slightly different manner, something that we call passive realization. In passivity control literature sign convention is such that energy dissipation has positive sign. With this convention, passive realization can be defined as follows.

Definition 6.2 (Passive Realization) A virtual environment is passively realized if for each admissible trajectory of the virtual interaction point the energy returned by the haptic simulation system to the user is always larger than the virtual energy generated by the virtual environment for the same trajectory.

In other words, during haptic interaction with a passively realized virtual environment, a user would experience a physically-realized force field that is always more dissipative than the force field specified by the virtual environment.

The concept of passive realization, as defined here, follows from the passivity analysis provided in [11], where two sufficient conditions for the passive realization of passive virtual environments can be found. We now extend this analysis to account for the effect of position measurement quantization error using most of the same notation and proofs outlines.

Note that the concept of passive realization, illustrated by Fig. 6.1, is stricter than the concept of passivity applied to haptic interaction. Passivity requires only a positive energy balance during interaction, counting on the haptic device intrinsic properties to dissipate some of the energy supplied by the user, irrespective of the properties of the virtual environment.

From the view point of control stability, passive realization is a stricter condition than passivity and, when applied to conservative force fields, ensures the stability of the interaction. In general, passively realized conservative virtual environment cannot be "stiffer" than what would be prescribed by a standard passivity analysis.

6.3.2 Passively Realized Virtual Environments

Given a haptic simulation system running with sampling period, T, where the haptic device, modeled as a damped multibody inertia subject to dry friction, has a damping matrix, $\mathbf{B}(t)$, and a dry friction matrix, $\mathbf{S}(t)$ (both positive definite), we desire to synthesize a nonlinear prescribed force field, $\boldsymbol{f}(t)$. The trajectory followed by the tip of the haptic device, $\boldsymbol{x}(t)$, at velocity, $\boldsymbol{v}(t)$, is discretized at times instants, $t_k = kT$, $\boldsymbol{x}_k = \boldsymbol{x}(t_k)$. We further assume that the position, $\boldsymbol{x}(t)$, is measured with encoders, resulting in a quantized position signal $\hat{\boldsymbol{x}}(t) = \boldsymbol{x}(t) - \Delta_{\boldsymbol{x}}(t)$, where $-\Delta_{\boldsymbol{x}}(t)$ represents the quantization error and where for $\tau \le 0$, $\boldsymbol{f}(\tau) = \boldsymbol{f}_0, \boldsymbol{x}(\tau) = \boldsymbol{x}_0$ and $\boldsymbol{v}(\tau) = 0$. We neglect the effects of output force quantization, finite register length, actuator saturation, and amplifier rise time. Calling $\hat{\boldsymbol{f}}(t) = \boldsymbol{f}(\hat{\boldsymbol{x}}(t))$ the force computed from the quantized position signal, with the above restrictions, the force returned by the haptic device

$$\hat{\boldsymbol{f}}_{\mathrm{v}}(\tau) = \sum_k \hat{\boldsymbol{f}}((k-n)T)p(\tau), \tag{6.2}$$

$$\text{where,} \quad p(\tau) = \begin{cases} 1, & \text{if } kT \le \tau < (k+1)T, \\ 0, & \text{otherwise,} \end{cases} \tag{6.3}$$

is a pulse. The error introduced by spatial quantization is

$$\Delta_{\boldsymbol{f}}(\tau) = \boldsymbol{f}_{\mathrm{v}}(\tau) - \hat{\boldsymbol{f}}_{\mathrm{v}}(\tau) \tag{6.4}$$

$$= \sum_k \Big(\boldsymbol{f}((k-n)T) - \hat{\boldsymbol{f}}((k-n)T)\Big)\, p(\tau), \tag{6.5}$$

$$\text{where,} \quad \boldsymbol{f}_{\mathrm{v}}(\tau) = \sum_k \boldsymbol{f}((k-n)T)p(\tau), \tag{6.6}$$

is the time-sampled force without quantization, see Fig. 6.2. Delay is represented by an integer number of samples, $n \in \mathbb{N}$, and $\Delta_{\boldsymbol{f}}(\tau)$ represents the error in the force introduced by the spatial quantization at the sample time t_k for $t_k \le \tau \le t_{k+1}$. The latter quantity can be very complex since it is a function of all the quantized position samples up to time τ, but it is constant during each sample interval.

We can rewrite Definition 6.2 to read:

Definition 6.3 (Passive Realization by device) If the haptic device can be assumed to behave like articulated rigid bodies subject to viscous and dry friction, a virtual

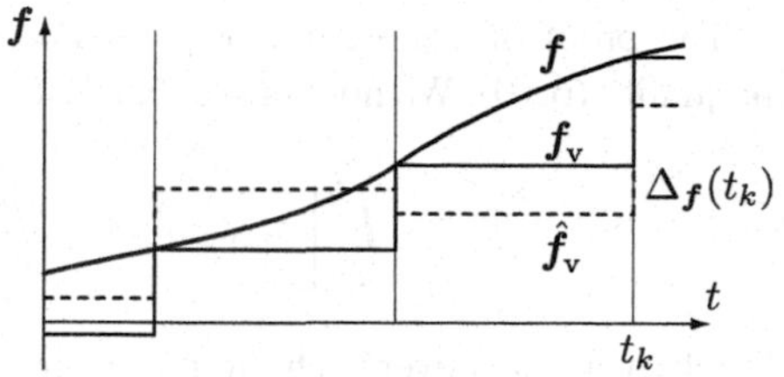

Fig. 6.2 The effects of quantization and zero-order hold sampling

force field $\boldsymbol{f}(t)$ is passively realized if

$$\int_0^t \left[\hat{\boldsymbol{f}}_{\text{v}}(\tau) - \boldsymbol{f}(\tau) + \mathbf{B}\boldsymbol{v}(\tau) + \mathbf{S}\frac{\boldsymbol{v}(\tau)}{|\boldsymbol{v}(\tau)|}\right]^{\top} \boldsymbol{v}(\tau)\mathrm{d}\tau \geq 0, \tag{6.7}$$

for all possible trajectories $\boldsymbol{v}(t)$.

We now introduce sufficient conditions for the passive realization of static virtual environments. There are many complex examples of static virtual environments, such as haptic textures [4].

Theorem 1 (Passive Realization of Static Virtual Environments) *Given the position of the haptic device $\boldsymbol{x}(t)$, its velocity $\boldsymbol{v}(t)$, and its acceleration $\boldsymbol{a}(t)$, a static force field $\boldsymbol{f}(\boldsymbol{x})$ is passively realized if $\forall t, t_1,$ and t_2, $t_1 < t_2$, $\exists \alpha, 0 \leq \alpha \leq 1$ and if*

1. $|\boldsymbol{f}(\boldsymbol{x}(t_2)) - \boldsymbol{f}(\boldsymbol{x}(t_1))| \leq \frac{\sigma_0(\mathbf{B})}{(n+1)T}|\boldsymbol{x}(t_2) - \boldsymbol{x}(t_1)|$,
2. $|\boldsymbol{a}(t)| \leq (1-\alpha)\frac{\sigma_0(\mathbf{S})}{\sigma_0(\mathbf{B})((n+1)T)}$,
3. *$\boldsymbol{v}$ is continuous, and*
4. $|\Delta_{\boldsymbol{f}}(t)| \leq \alpha\sigma_0(\mathbf{S})$,

where $\sigma_0(\mathbf{B})$ and $\sigma_0(\mathbf{S})$ are the smallest singular values over the workspace of the damping and friction matrix respectively.

Proof To prove (6.7) consider the effect of quantization and sample-and-hold,

$$\hat{\boldsymbol{f}}_{\text{v}}(\tau) = \boldsymbol{f}_{\text{v}}(\tau) - \Delta_{\boldsymbol{f}}(\tau), \tag{6.8}$$

and separate the contribution of dry friction. By showing that

$$\int_0^t \left[-\Delta_{\boldsymbol{f}}(\tau) + \mathbf{S}\frac{\alpha\boldsymbol{v}(\tau)}{|\boldsymbol{v}(\tau)|}\right]^{\top} \boldsymbol{v}(\tau)\mathrm{d}\tau \geq 0, \tag{6.9}$$

and that

$$\int_0^t \left[\boldsymbol{f}_{\text{v}}(\tau) - \boldsymbol{f}(\tau) + \mathbf{B}\boldsymbol{v}(\tau) + \mathbf{S}\frac{(1-\alpha)\boldsymbol{v}(\tau)}{|\boldsymbol{v}(\tau)|}\right]^{\top} \boldsymbol{v}(\tau)\mathrm{d}\tau \geq 0, \tag{6.10}$$

then, adding the inequalities (6.9) and (6.10) yields (6.7).

The proof of Theorem 1 in [11] shows that Conditions 1–2 imply the second inequality (6.10). We now prove that $\forall t$ and $t_k \le t < t_{k+1}$,

$$\int_{t_k}^{t} \left[-\Delta_f(\tau) + \mathbf{S}\frac{\alpha \boldsymbol{v}(\tau)}{|\boldsymbol{v}(\tau)|} \right]^{\top} \boldsymbol{v}(\tau)\mathrm{d}\tau \ge 0, \tag{6.11}$$

which means that over each sampling period the friction dissipates more energy than the spurious energy introduced by the quantization error. In fact,

$$\int_{t_k}^{t} \left[\mathbf{S}\frac{\boldsymbol{v}(\tau)}{|\boldsymbol{v}(\tau)|} \right]^{\top} \boldsymbol{v}(\tau)\mathrm{d}\tau = \int_{t_k}^{t} \frac{\boldsymbol{v}(\tau)^{\top}\mathbf{S}^{\top}(\tau)\boldsymbol{v}(\tau)}{|\boldsymbol{v}(\tau)|}\mathrm{d}\tau \tag{6.12}$$

$$\ge \sigma_0(\mathbf{S}) \int_{t_k}^{t} |\boldsymbol{v}(\tau)|\mathrm{d}\tau \tag{6.13}$$

$$\ge \sigma_0(\mathbf{S})|\boldsymbol{x}(t) - \boldsymbol{x}_k|, \tag{6.14}$$

where the integral in (6.13) is the length of the trajectory of the haptic device between t_k and t . Since the shortest distance between two points in Euclidean space is a straight line, we can find a lower bound for this integral by setting the velocity at a constant speed $\boldsymbol{v}^* = -(\boldsymbol{x}(t) - \boldsymbol{x}_k)/(t - t_k)$. We then note that, for $t_k \le t < t_{k+1}$, $\Delta_f(t) = \Delta_f(t_{k-n})$ is the constant quantization error at the time step $k - n$. The integral in (6.11) is independent from the velocity and can be evaluated for $\boldsymbol{v}(\tau)$:

$$\int_{t_k}^{t} \left[\mathbf{S}\frac{\alpha \boldsymbol{v}(\tau)}{|\boldsymbol{v}(\tau)|} \right]^{\top} \boldsymbol{v}(\tau)\mathrm{d}\tau \ge \alpha\sigma_0(\mathbf{S})|\boldsymbol{x}(t) - \boldsymbol{x}_k|. \tag{6.15}$$

We now evaluate

$$\int_{t_k}^{t} \left[\mathbf{S}\frac{\alpha \boldsymbol{v}(\tau)}{|\boldsymbol{v}(\tau)|} \right]^{\top} \boldsymbol{v}(\tau)\mathrm{d}\tau \tag{6.16}$$

$$\ge \sigma_0(\mathbf{S})|\boldsymbol{x}(t) - \boldsymbol{x}_k| \tag{6.17}$$

$$\ge |\Delta_f(t_{k-n})||\boldsymbol{x}(t) - \boldsymbol{x}_k| \tag{6.18}$$

$$\ge \Delta_f(t_{k-n})^{\top}(\boldsymbol{x}_k - \boldsymbol{x}(t)) \tag{6.19}$$

$$= \int_{t_k}^{t} \Delta_f(t_{k-n})^{\top}\frac{\boldsymbol{x}_k - \boldsymbol{x}(t)}{t - t_k}\mathrm{d}\tau \tag{6.20}$$

$$= \int_{t_k}^{t} \Delta_f(\tau)^{\top}\boldsymbol{v}(\tau)\mathrm{d}\tau, \tag{6.21}$$

which can be rearranged to become

$$\int_{t_k}^{t} \left[\mathbf{S}\frac{\alpha \boldsymbol{v}(\tau)}{|\boldsymbol{v}(\tau)|} - \Delta_f(\tau) \right]^{\top} \boldsymbol{v}(\tau)\mathrm{d}\tau \ge 0. \tag{6.22}$$

Hence, from (6.9),

$$\int_0^t \left[\mathbf{S}\frac{\alpha \boldsymbol{v}(\tau)}{|\boldsymbol{v}(\tau)|} - \Delta_f(\tau)\right]^\top \boldsymbol{v}(\tau)\mathrm{d}\tau \tag{6.23}$$

$$= \int_{t_k}^t \left[\mathbf{S}\frac{\alpha \boldsymbol{v}(\tau)}{|\boldsymbol{v}(\tau)|} - \Delta_f(\tau)\right]^\top \boldsymbol{v}(\tau)\mathrm{d}\tau \tag{6.24}$$

$$+ \sum_{i=0}^{k-1} \int_{t_i}^{t_{i+1}} \left[\mathbf{S}\frac{\alpha \boldsymbol{v}(\tau)}{|\boldsymbol{v}(\tau)|} - \Delta_f(\tau)\right]^\top \boldsymbol{v}(\tau)\mathrm{d}\tau \geq 0. \tag{6.25}$$

□

6.4 Examples and Discussion

The newly introduced Condition 4 can be rewritten for static force fields of the form $\boldsymbol{f}(t) = \boldsymbol{f}(\boldsymbol{x}(t))$. If

$$\Delta_f(\boldsymbol{x}(\tau)) = [\boldsymbol{f}_\mathrm{v}(\boldsymbol{x}(\tau)) - \boldsymbol{f}_\mathrm{v}(\boldsymbol{x}(\tau) - \Delta_x(\tau))], \tag{6.26}$$

then we can rewrite Condition 4 as

$$\forall\tau, \quad |\boldsymbol{f}_\mathrm{v}(\boldsymbol{x}(\tau)) - \boldsymbol{f}_\mathrm{v}(\boldsymbol{x}(\tau) - \Delta_x(\tau))| \leq \alpha\sigma_0(\mathbf{S}), \tag{6.27}$$

which is equivalent to imposing a limit on the force error at any time τ and position $\boldsymbol{x}(\tau)$. Linearizing $\boldsymbol{f}$ around the point $\boldsymbol{x}(t)$ and employing the Jacobian of the force field, $\mathbf{J}_f$, gives

$$|\boldsymbol{f}_\mathrm{v}(\boldsymbol{x}(\tau)) - \boldsymbol{f}_\mathrm{v}(\boldsymbol{x}(\tau) - \Delta_x(\tau))| \tag{6.28}$$

$$= |\boldsymbol{f}(\boldsymbol{x}_k) - \boldsymbol{f}(\boldsymbol{x}_k - \Delta_x(\tau))| \tag{6.29}$$

$$\leq |\mathbf{J}_f|| - \Delta_x(\tau)| \leq |\mathbf{J}_f|\Delta, \tag{6.30}$$

where

$$\Delta \triangleq \max_{x \in \text{workspace}} |\Delta_x| \tag{6.31}$$

is the greatest distance by which a user can move the haptic device before triggering an encoder pulse anywhere in the workspace and where

$$|\mathbf{J}_f| \triangleq \max_{x \in [\boldsymbol{x}_k - \Delta_x(\tau), \boldsymbol{x}_k]} |\mathbf{J}_f(\boldsymbol{x})| \tag{6.32}$$

is the steepest "displacement-force gain" in the interval $[\boldsymbol{x}_k - \Delta_x(\tau), \boldsymbol{x}_k]$. This yields an equivalent condition for Condition 4 in the case of linear virtual environments,

$$\sigma_n(\mathbf{J}_f) \leq \alpha\frac{\sigma_0(\mathbf{S})}{\Delta}, \tag{6.33}$$

where $\sigma_n(\mathbf{J}_f)$ is the largest singular value of the Jacobian matrix of the force field over the admissible workspace.

For a linearized static virtual environment, we can also rewrite Condition 1 to read

$$\sigma_n(\mathbf{J}_f) \leq \frac{\sigma_0(\mathbf{B})}{(n+1)T}. \tag{6.34}$$

Where the Jacobian matrix of the force field is not defined, as for example on the boundary of a virtual wall, we would compute the maximum norm of the directional derivatives of the force field.

6.4.1 *Linearized Virtual Environments*

6.4.1.1 Conservative Case, No Delay

An interesting case arises when considering static affine virtual environments of the form $\boldsymbol{f}(\boldsymbol{x}) = -\mathbf{K}\boldsymbol{x} + \boldsymbol{f}_0$ when there is no delay, for instance obtained by linearizing a nonlinear field around an operating point. In this specific case, it is possible to derive a tighter condition for the passive realization of the virtual environment. By comparing the new condition with the passivity analysis of virtual walls, the concept of passive realization is further specialized.

Proposition 1 *If the force field* $\boldsymbol{f}(\boldsymbol{x}) = -\mathbf{K}\boldsymbol{x} + \boldsymbol{f}_0$ *is conservative, the sufficient conditions for a passively realized environment are*

1. $\lambda_n(\mathbf{K}) \leq \frac{2\sigma_0(\mathbf{B})}{T}$,
2. $\boldsymbol{v}$ *is continuous, and*
3. $\sigma_n(\mathbf{K}) \leq \frac{\sigma_0(\mathbf{S})}{\Delta}$,

where $\lambda_n(\mathbf{K})$ *is the largest eigenvalue of* $\mathbf{K}$, *which has only real values (because it is symmetric, since the force field is conservative).*

Proof We show that the simplified Conditions 1, 2, 3 are sufficient for Eq. (6.7). We first show that Condition 1 and 3 are sufficient conditions if spatial quantization can be neglected. The energy balance over the time period depends only on the endpoints, because the force field is conservative. As a result, we may evaluate the integral over a path traversed at constant speed:

$$-\int_{t_k}^{t} \left[\boldsymbol{f}_{\mathrm{v}}(\tau) - \boldsymbol{f}(\tau)\right]^{\top} \boldsymbol{v}(\tau)\mathrm{d}\tau \tag{6.35}$$

$$= -\int_{t_k}^{t} \left[\boldsymbol{f}(t_k) - \boldsymbol{f}(\tau)\right]^{\top} \boldsymbol{v}(\tau)\mathrm{d}\tau \tag{6.36}$$

$$= \int_{t_k}^{t} \left[\mathbf{K}\boldsymbol{x}_k - \mathbf{K}\boldsymbol{x}(\tau)\right]^{\top} \boldsymbol{v}(\tau)\mathrm{d}\tau \tag{6.37}$$

$$= \int_{t_k}^{t} \left[\mathbf{K}(\boldsymbol{x}_k + (\tau - t_k)\boldsymbol{v}(t) - \boldsymbol{x}_k)\right]^{\top} \boldsymbol{v}(\tau)\mathrm{d}\tau \tag{6.38}$$

$$= \int_{t_k}^{t} [\mathbf{K}(\tau - t_k)\boldsymbol{v}(t)]^\top \boldsymbol{v}(\tau)\mathrm{d}\tau \tag{6.39}$$

$$= \frac{(\boldsymbol{x}_k - \boldsymbol{x}(t))^\top \mathbf{K}^\top (\boldsymbol{x}_k - \boldsymbol{x}(t))}{2}. \tag{6.40}$$

Call $\boldsymbol{u}$ a unit vector aligned with $(\boldsymbol{x}(t_k) - \boldsymbol{x}(t))$, the above can be rewritten

$$\boldsymbol{u}^\top \mathbf{K}^\top \boldsymbol{u} \frac{|\boldsymbol{x}(t) - \boldsymbol{x}(t_k)|^2}{2} \tag{6.41}$$

$$\leq \lambda_n(\mathbf{K}) \frac{|\boldsymbol{x}(t) - \boldsymbol{x}(t_k)|^2}{2} \tag{6.42}$$

$$\leq \sigma_0(\mathbf{B}) \frac{|\boldsymbol{x}(t) - \boldsymbol{x}(t_k)|^2}{T}. \tag{6.43}$$

The trajectory length is minimized when traversed at the constant velocity, $\boldsymbol{v}^* = -(\boldsymbol{x}(t) - \boldsymbol{x}_k)/(t - t_k)$. Therefore, the integral $\| \int_{t_k}^{t} \boldsymbol{v}(\tau)^2 \mathrm{d}\tau \|$ admits a lower bound given by

$$\sigma_0(\mathbf{B}) \frac{|\boldsymbol{x}(t) - \boldsymbol{x}_k|^2}{T} \leq \frac{\sigma_0(\mathbf{B})}{T} \left| \int_{t_k}^{t} \boldsymbol{v}(\tau)\mathrm{d}\tau \right|^2 \tag{6.44}$$

$$\leq \sigma_0(\mathbf{B}) \int_{t_k}^{t} |\boldsymbol{v}(\tau)|^2 \mathrm{d}\tau \leq \int_{t_k}^{t} \boldsymbol{v}(t)^\top \mathbf{B}^\top \boldsymbol{v}(t)\mathrm{d}\tau \tag{6.45}$$

applying the Cauchy-Schwarz inequality. From (6.35)–(6.45) we conclude that

$$- \int_{t_k}^{t} \left[\boldsymbol{f}_{\mathrm{v}}(\tau) - \boldsymbol{f}(\tau)\right]^\top \boldsymbol{v}(\tau)\mathrm{d}\tau \leq \int_{t_k}^{t} \boldsymbol{v}(t)^\top \mathbf{B}^\top \boldsymbol{v}(t)\mathrm{d}\tau, \tag{6.46}$$

or,

$$\int_{t_k}^{t} \left[\boldsymbol{f}_{\mathrm{v}}(\tau) - \boldsymbol{f}(\tau) + \mathbf{B}\boldsymbol{v}(\tau)\right]^\top \boldsymbol{v}(\tau)\mathrm{d}\tau \geq 0. \tag{6.47}$$

To complete the proof we must show that

$$\int_{0}^{t} \left[\mathbf{S}\frac{\boldsymbol{v}(\tau)}{|\boldsymbol{v}(\tau)|} - \Delta_f(\tau)\right]^\top \boldsymbol{v}(\tau)\mathrm{d}\tau \geq 0. \tag{6.48}$$

Equations (6.26)–(6.33) give

$$\sigma_n(\mathbf{J}_f) \leq \frac{\sigma_0(\mathbf{S})}{\Delta}, \tag{6.49}$$

which implies that

$$|\Delta_f(t)| \leq \sigma_0(\mathbf{S}). \tag{6.50}$$

The proof then follows from (6.12)–(6.22). □

In particular, for a virtual wall we have

$$\sigma_n(\mathbf{K}) = \lambda_n(\mathbf{K}) = \begin{cases} K_V, & \text{inside,} \\ 0, & \text{outside.} \end{cases} \tag{6.51}$$

Condition 1 is in complete agreement with previous studies of passivity for virtual walls, for example, with Colgate's equation [5]. On the other hand, Condition 3 does not agree exactly with the result in [1], namely, that $K_V \leq 2\sigma_0(\mathbf{S})/\Delta$ is a necessary and sufficient condition for passivity. This discrepancy is due to the fact that passivity limits only the flow of energy between the user and the haptic device but does not restrict the excess energy with respect to the virtual environment. In fact, during a compression phase, the energy stored in a virtual wall could become greater than the energy provided by the user while pushing in, but passivity is guaranteed as long as this extra energy is dissipated during the release phase. This argument does not apply to passively realized virtual environments, because at any time, t, the energy in the virtual environment must be smaller than the energy supplied by the user.

Remark 1 The factor 2 introduced in Condition 1 is specific to linear, conservative, non-delayed virtual environments. In all other cases, the more restrictive sufficient condition stated in Theorem 1 must be used.

Remark 2 In the best-case scenario, i.e. when spatial quantization is negligible, Condition 1 and 3 are sufficient for the passive realization of a conservative force field. As a consequence, if $\mathbf{K}$ is negative definite, the environment is always passively realizable. For example, a positive spring synthesized according to $\boldsymbol{f} = K_V \boldsymbol{x}$ has such property for $K_V \geq 0$.

6.4.1.2 Nonconservative Case, No Delay

We now consider affine force fields of the form $\boldsymbol{f}(\boldsymbol{x}) = -\mathbf{K}\boldsymbol{x} + \boldsymbol{f}_0$ with $\mathbf{K}^\top \neq \mathbf{K}$. The sufficient conditions follow directly from Theorem 1:

1. $\sigma_n(\mathbf{K}) \leq \frac{\sigma_0(\mathbf{B})}{T}$,
2. $|\boldsymbol{a}(t)| \leq (1-\alpha)\frac{\sigma_0(\mathbf{S})}{T\sigma_0(\mathbf{B})}$,
3. $\sigma_n(\mathbf{K}) \leq \alpha\frac{\sigma_0(\mathbf{S})}{\Delta}$.

The major difference between conservative and non conservative field is the effect of the trajectory on the energy generated by the virtual environment; in fact, in the conservative case it is possible to choose a convenient integration path, while in the non conservative case the worst case trajectory must be considered.

Condition 1 for non conservative case differs from the conservative case as it requires the largest singular and a factor of 1, instead of a factor of 2. In addition, Conditions 2 and 3 are related. If Condition 3 is met and if

$$\alpha = \sigma_n(\mathbf{K})\frac{\Delta}{\sigma_0(\mathbf{S})} \leq 1, \tag{6.52}$$

then Condition 2 becomes

$$|\boldsymbol{a}(t)| \leq \frac{\sigma_0(\mathbf{S})}{T\sigma_0(\mathbf{B})} - \sigma_n(\mathbf{K})\frac{\Delta}{T\sigma_0(\mathbf{B})} \leq \frac{\sigma_0(\mathbf{S})}{T\sigma_0(\mathbf{B})} - \frac{\Delta}{T^2} \tag{6.53}$$

by applying Condition 1. This expression shows that a coarser spatial quantization, i.e. a larger Δ, limits the trajectories along which a nonconservative virtual environment is guaranteed to be passively realized.

6.4.1.3 Delayed

In the general case of a delayed linear environment the combined Conditions 1 and 2 as per (6.53) become

$$|\boldsymbol{a}(t)| \leq \frac{\sigma_0(\mathbf{S})}{((n+1)T)\sigma_0(\mathbf{B})} - \frac{\Delta}{((n+1)T)^2} \tag{6.54}$$

which shows that when delay increases, the effect of the quantization over the admissible trajectories decreases quickly since the range of passively realizable stiffnesses is greatly reduced by Condition 1.

6.4.2 Experimental Results

To illustrate the difference between the notions of passivity and passive realization we recorded the interaction of a user with two-dimensional, linear, static, nonconservative virtual environments. The experimental set-up was in all practical aspects identical to that described in Chaps. 3 and 4. Its description is not repeated here.

For each case, generative and dissipative trajectories were captured. These two examples show the effect of time discretization when the actuator command is reconstructed using a zero-order hold. Data were logged at approximately 9 kHz (which is 10 times the mechanical bandwidth of the haptic device) and the resolution was approximately 15 μm. The virtual environments were computed at 9 Hz, with no delay, hence the lack of passive realization can only be due to low temporal resolution and zero-order reconstruction.

It is important to notice that the energy exchanged with the user is computed and not measured. The plots do not account for the dissipation in the haptic device, which could affect the energy gap. The point of the two examples is to illustrate how passive realization can be used to analyze both conservative and not conservative virtual environments.

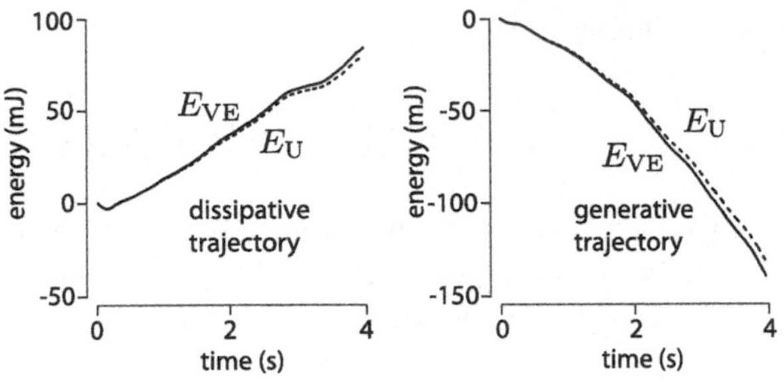

Fig. 6.3 Energy balance when a user explores a vortex-like virtual environment with dissipative and generative trajectories. E_{VE} is the virtual energy and E_U is the energy exchanged with the user

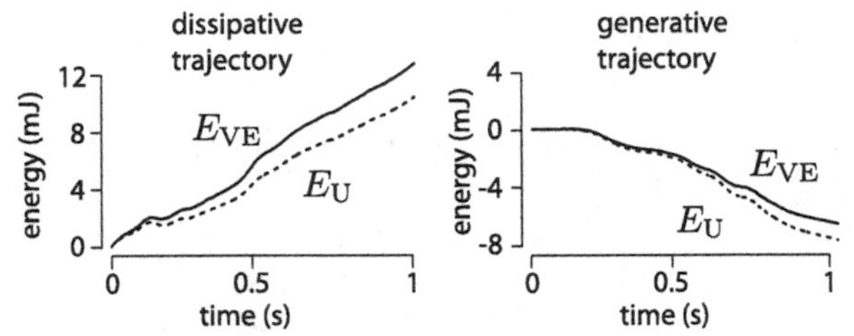

Fig. 6.4 Energy balance when a user explores a iris-like virtual environment with dissipative and generative trajectories. E_{VE} is the virtual energy and E_U is the energy exchanged with the user

6.4.2.1 Vortex

The first example is a vortex-like force field described by

$$f(\boldsymbol{x}(t)) = K \begin{pmatrix} y \\ -x \end{pmatrix}, \tag{6.55}$$

where $K = 0.05$ N/mm. Figure 6.3 shows the energy balance during interaction. When the user works against the force field of the vortex along a dissipative trajectory, the resulting interaction is not passively realized; but it is passively realized when moving along the force field. According to standard passivity theory, while the dissipative case would be acceptable, the generative case wouldn't, since the simulation supplies large amounts of energy to the user.

6.4.2.2 Iris

The second virtual environment is a circular cavity with a superimposed tangential component (like a camera iris).

$$f(\boldsymbol{x}(t)) = \begin{cases} \left[K_1\binom{-y}{x} - K_2\binom{x}{y}\right]\frac{\|\boldsymbol{x}\|-R}{\|\boldsymbol{x}\|}, & \text{if } \left\|\binom{x}{y}\right\| > R, \\ 0, & \text{otherwise.} \end{cases} \tag{6.56}$$

In this example, see Fig. 6.4, neither of the two exploration modes are passively realized. Again, the haptic exploration along the dissipative trajectory is passive while the generative interaction is not.

These examples show that standard passivity theory cannot be used to analyze the energy behavior of these two examples of complex force fields, because the theory does not consider their energy characteristic. Other approaches however could be used to enforce a correct energy balance; for example a passivity observer could be adapted to track the energy imbalance between the virtual environment and the real interaction so that the passivity controller could introduce virtual dissipation when needed. Similarly, port-Hamiltonian approaches could be adapted to enforce the passive realization of nonlinear force fields. Nevertheless, these online methods are blind to the properties of the virtual environment and could not be used as tools to analyse the properties different synthesis algorithms.

6.5 Conclusion and Future Work

In this chapter we introduced the concept of passively realized virtual environments, which is intimately related to passivity, and provides a unambiguous tool to analyze the energetic properties of virtual force fields. We also adapted previous results from passivity analysis and extended them to analyze, for the first time, the effect of spatial quantization on the energy behavior of realized nonlinear, multidimensional static virtual environments. Spatial quantization was shown to reduce the range of admissible trajectories along which a nonlinear static virtual environment can be passively realized. The results also confirms the usefulness of the Jacobian matrix of the force field as a tool to measure the "effective stiffness" of nonlinear virtual environments, as introduced in Chap. 4. Finally, we believe that the concept of passive realization so far studied for the case of static or slow varying fields could be extended to fast, time-varying force fields.

When both passivity and passive realization can be used to analyze the same haptic interaction, the newly introduce passive realization guarantees passivity but not the viceversa; i.e. there are passive haptic interactions that are not passively realized, thus passive realization is a more general framework but it forces more restrictions on conservative force fields.

Acknowledgements This work was funded by a Collaborative Research and Development Grant "High Fidelity Surgical Simulation" from the Natural Sciences and Engineering Council of Canada (NSERC), and by Immersion Corp. Additional funding is a from a Discovery Grant from NSERC for the second author. The authors would like to thanks the reviewers for several insightful comments.

References

1. Abbott, J.J., Okamura, A.M.: Effects of position quantization and sampling rate on virtual wall passivity. IEEE Trans. Robot. **21**(5), 952–964 (2005)
2. Adams, R.J., Hannaford, B.: Stable haptic interaction with virtual environments. IEEE Trans. Robot. Autom. **15**(3), 465–474 (1999)
3. Campion, G., Hayward, V.: Fundamental limits in the rendering of virtual haptic textures. In: Proceedings of the First Joint Eurohaptics Conference and Symposium on Haptic Interfaces for Virtual Environments and Teleoperator Systems, WHC'05, pp. 263–270 (2005)

4. Campion, G., Hayward, V.: On the synthesis of haptic textures. IEEE Trans. Robot. **24**(3), 527–536 (2008)
5. Colgate, J.E., Schenkel, G.G.: Passivity of a class of sampled-data systems: Application to haptic interfaces. J. Robot. Syst. **14**(1), 37–47 (1997)
6. Diolaiti, N., Niemeyer, G., Barbagli, F., Salisbury, J.K.: Stability of haptic rendering: Discretization, quantization, time delay, and coulomb effects. IEEE Trans. Robot. **22**(2), 256–268 (2006)
7. Gil, J.J., Sanchez, E., Hulin, T., Preusche, C., Hirzinger, G.: Stability boundary for haptic rendering: Influence of damping and delay. In: Proceedings of the IEEE International Conference on Robotics and Automation, pp. 124–129 (2007)
8. Hulin, T., Preusche, C., Hirzinger, G.: Stability boundary for haptic rendering: Influence of physical damping. In: Intelligent Robots and Systems, 2006 IEEE/RSJ International Conference on, pp. 1570–1575 (2006)
9. Lee, K., Lee, D.Y.: Adjusting output-limiter for stable haptic interaction with deformable objects. IEEE Trans. Control Syst. Technol. **17**(3), 768–779 (2009)
10. Lozano, R., Maschke, B., Brogliato, B., Egeland, O.: Dissipative Systems Analysis and Control: Theory and Applications. Springer, New York (2000)
11. Mahvash, M., Hayward, V.: High fidelity passive force reflecting virtual environments. IEEE Trans. Robot. **21**(1), 38–46 (2005)
12. Miller, B.E., Colgate, J.E., Freeman, R.A.: Guaranteed stability of haptic systems with nonlinear virtual environments. IEEE Trans. Robot. Autom. **16**(6), 712–719 (2000)
13. Ryu, J.-H., SangKim, Y., Hannaford, B.: Sampled- and continuous-time passivity and stability of virtual environments. IEEE Trans. Robot. **20**, 772–776 (2004)
14. Stramigioli, S., Secchi, C., van der Schaft, A.J., Fantuzzi, C.: Sampled data systems passivity and discrete port-hamiltonian systems. IEEE Trans. Robot. **21**(4), 574–587 (2005)

Chapter 7
Texturing Curved Surfaces

Abstract This chapter discusses the extension of the characteristic number of virtual haptic textures to the case of non flat surfaces. This extension shows that there is an effect of surface curvature on the passivity of haptics textures, but that there are other effects and artifacts that cannot be captured by the characteristic number. The chapter is concluded by examples of the characteristic number of two texture algorithms applied to a cilindrical curved surface.

7.1 Introduction

In this Chapter, the concept of characteristic number of an algorithm is extended to curved surfaces in 3D. This number makes it possible to evaluate the passivity of haptic synthesis when haptic textures are mapped over curved surfaces. We also analyze newly discovered artifacts which result from the projection of spatial trajectories onto curved surfaces.

The function that maps the 3D surface to the texture space plays a central role in the change in closed loop behavior of haptic simulation as it affects directly the characteristic number and thus the passivity margin of the haptic interaction. The mapping of planar patches onto curved surfaces is a widely explored topic in the computer graphics domain [5], and the same techniques can be applied to generate the mapping function for haptic textures. For this reason, the texture map problem is not discussed in detail in this chapter.

The extensions contained in this chapter have not been experimentally validated, but offer interesting insights on the problems related to extending passivity analysis to surfaces with complex shapes. There are however artifacts and sources of instability that cannot be captured by the characteristic number; for examples all the effects due to the global geometry of the surface evade the analysis carried out in this chapter.

7.1.1 3D Surfaces

Given $z = \mathbf{S}(x, y)$ a regular 3D surface and $h(i, j)$ a height field mapped on the same surface, we would like to express the characteristic number for a given texture

G. Campion, *The Synthesis of Three Dimensional Haptic Textures: Geometry, Control, and Psychophysics*, Springer Series on Touch and Haptic Systems,
DOI 10.1007/978-0-85729-576-7_7,

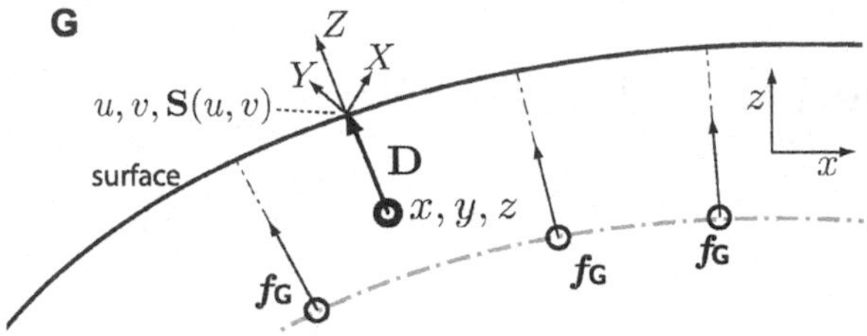

Fig. 7.1 The god-object algorithm applied to a curved surface. This figure shows a vertical x, z slice of the generic 3D rendering problem. The *dashed line* represents a line of constant deflection inside a convex surface, the displacement of the interaction point on this line forces a longer displacement of the god-object point on the surface **S**. The normal coordinate frame is on the surface **S**

algorithm. It turns out that the curvature of **S** is the factor that drives the passivity margins of the simulation.

In the rest of this Chapter, we assume that the surface **S** is regular in the parametrization $(x, y, \mathbf{S}(x, y))$, that is, the map $z = \mathbf{S}(x, y)$ is differentiable (all derivatives are continuous) and its inverse map $\mathbf{S}^{-1}$ is continuous.

7.2 The God-Object

The first step is to analyze the properties of the algorithm used to render the untextured surface $z \leq \mathbf{S}(x, y)$. A penalty-based method similar to the god-object method, as described in [6], is used to generate a force field normal to the surface.

Figure 7.1 illustrates the god-object algorithm for the 3D case. The virtual interaction point is at position

$$\mathbf{x} = [x \quad y \quad z]^\top. \tag{7.1}$$

It is then possible to compute two functions $u = u(\mathbf{x})$ and $v = v(\mathbf{x})$ such that a point on the surface (called 'god-object' point)

$$\mathbf{s} = \mathbf{s}(\mathbf{x}) = [u \quad v \quad \mathbf{S}(u, v)]^\top \tag{7.2}$$

minimizes the distance, $\|\mathbf{D}\|$, between the virtual interaction point and the surface, where

$$\mathbf{D}(\mathbf{x}) = \mathbf{s} - \mathbf{x} = \begin{bmatrix} u - x \\ v - y \\ \mathbf{S}(u, v) - z \end{bmatrix}. \tag{7.3}$$

To simplify the notation, let's abbreviate the partial derivatives of a quantity **A** as $\frac{\partial \mathbf{A}}{\partial b} = \mathbf{A}_b$.

The distance $\|\mathbf{D}\|$ is minimized when

$$\mathbf{D}_u{}^\top \mathbf{D} = u - x + \mathbf{S}_u(u, v)(\mathbf{S}(u, v) - z) = 0, \tag{7.4}$$

$$\mathbf{D}_v{}^\top \mathbf{D} = v - y + \mathbf{S}_v(u, v)(\mathbf{S}(u, v) - z) = 0, \tag{7.5}$$

giving

$$\mathbf{S}_u(u, v) = \frac{u - x}{\mathbf{S}(u, v) - z}, \tag{7.6}$$

$$\mathbf{S}_v(u, v) = \frac{v - y}{\mathbf{S}(u, v) - z}, \tag{7.7}$$

that is, the partial derivatives of the surface function with respect to its parametrization. Following the original god-object approach, an energy function is given by

$$E(\mathbf{x}) = -\frac{K_\mathrm{V}}{2}\mathbf{D}(\mathbf{x})^\top \mathbf{D}(\mathbf{x}), \tag{7.8}$$

such that the interaction force f_D is computed by differentiating $E(\mathbf{x})$ with respect to $\mathbf{x}$.

$$\begin{aligned} f_\mathsf{D}(\mathbf{x}) &= \begin{bmatrix} E_x & E_y & E_z \end{bmatrix}^\top \\ &= K_\mathrm{V}(\mathbf{s} - \mathbf{x}) = K_\mathrm{V}\mathbf{D}(\mathbf{x}). \end{aligned} \tag{7.9}$$

Finally, the Jacobian matrix $\mathbf{J}_{f_\mathsf{D}}(\mathbf{x})$ of $f_\mathsf{D}(\mathbf{x})$ becomes

$$\mathbf{J}_{f_\mathsf{D}}(\mathbf{x}) = K_\mathrm{V}\begin{bmatrix} u_x(\mathbf{x}) - 1 & u_y(\mathbf{x}) & u_z(\mathbf{x}) \\ v_x(\mathbf{x}) & v_y(\mathbf{x}) - 1 & v_z(\mathbf{x}) \\ \mathbf{S}_x(u, v) & \mathbf{S}_y(u, v) & \mathbf{S}_z(u, v) - 1 \end{bmatrix}. \tag{7.10}$$

7.2.1 Change of Coordinates

A change of coordinates can simplify the computation of the derivatives in the Jacobian matrix. Given a regular surface $\mathbf{S}$ in $\mathbb{R}^3$, there exists a parameterization in a reference frame X, Y, Z (called normal reference frame) such that the surface can be written

$$Z = \overline{\mathbf{S}}(X, Y), \tag{7.11}$$

with the property that the first order the second order mixed partial derivatives vanish for $X = Y = 0$:

$$\overline{\mathbf{S}}(0,0) = \overline{\mathbf{S}}_X(0,0) = \overline{\mathbf{S}}_Y(0,0) = \overline{\mathbf{S}}_{XY}(0,0) = \overline{\mathbf{S}}_{YX}(0,0) = 0. \tag{7.12}$$

If the point $(u, v, \mathbf{S}(u, v))$ is not umbilical, the axes X, Y can be aligned along the directions of maximum and minimum curvature [1]. The orientation of axis Z is chosen to ensure that the points $[0\ 0\ Z]^\top$ are inside the surface for $Z \leq 0$. In the new parametrization, from Eqs. (7.6), (7.7), and (7.12) it is possible to show that inside the surface,

$$U(0,0,Z)=0, \tag{7.13}$$

$$V(0,0,Z)=0, \tag{7.14}$$

$$\|\overline{\mathbf{D}}(0,0,Z)\|=-Z, \tag{7.15}$$

where $\overline{\mathbf{D}}$ is the penetration vector with respect to the new reference frame, and $\|\mathbf{D}\|=\|\overline{\mathbf{D}}\|$. Scalar quantities referred to this new reference frame are capitalized, while vectors and matrices are underlined.

This change of coordinates rotates the vector $\mathbf{D}$ to coincide with Z axis, so that motions normal to the surface in the initial reference frame are now along the vertical axis. Since the movements of the virtual interaction point along the direction of vector $\mathbf{D}$ leave the god-object location stationary in (u, v) coordinates, it is possible to evaluate the Jacobian matrix of the force field over the entire Z axis in the normal reference frame.

The change of coordinates depends only on the position of the god object, therefore, the analysis of the Jacobian matrix can be restricted to the Z axis ($\overline{\mathbf{J}}_{f_\mathbf{D}}(0,0,Z)$). The characteristic number computed in this new reference frame can be expressed in terms of invariant elements of the surface, such as curvature. In the normal coordinate frame (X, Y, Z) the Jacobian becomes

$$\overline{\mathbf{J}}_{f_\mathbf{D}}(0,0,Z)=K_\mathrm{V}\begin{bmatrix} U_X(0,0,Z)-1 & U_Y(0,0,Z) & U_Z(0,0,Z) \\ V_X(0,0,Z) & V_Y(0,0,Z)-1 & V_Z(0,0,Z) \\ 0 & 0 & -1 \end{bmatrix}. \tag{7.16}$$

The computation of $U_{X,Y,Z}(0,0,Z)$ and $V_{X,Y,Z}(0,0,Z)$ begins by differentiating the Eq. (7.5) with respect to $\mathbf{X}$ and by virtue of the properties in Eq. (7.12) we can write

$$\begin{aligned}
\frac{\partial}{\partial X}\overline{\mathbf{D}}_U{}^{\top}\overline{\mathbf{D}} &= 0 \\
&= \frac{\partial}{\partial X}\Big(U - X + \overline{\mathbf{S}}_U\cdot(\overline{\mathbf{S}}-Z)\Big) \\
&= U_X - 1 + \overline{\mathbf{S}}_{UX}\cdot(-Z) + \overline{\mathbf{S}}_U\cdot\overline{\mathbf{S}}_X \\
&= U_X - 1 + \overline{\mathbf{S}}_{UU}\cdot U_X\cdot(-Z) + \overline{\mathbf{S}}_{UV}\cdot V_X\cdot(-Z) \\
&= U_X - 1 + \overline{\mathbf{S}}_{UU}\cdot U_X\cdot(-Z) = 0
\end{aligned}$$

hence

$$U_X(0,0,Z)=\Big(1-\overline{\mathbf{S}}_{UU}(0,0)\cdot Z\Big)^{-1}. \tag{7.17}$$

Similarly, it can be shown that $U_Y(0,0,Z)=U_Z(0,0,Z)=V_X(0,0,Z)=V_Z(0,0,Z)=0$ and $V_Y(0,0,Z)=(1-\overline{\mathbf{S}}_{VV}(0,0)\cdot Z)^{-1}$.

The Jacobian matrix is then greatly simplified and now assumes a diagonal structure

$$\bar{\mathbf{J}}_{f_\mathsf{D}}(Z) = K_\mathrm{V} \begin{bmatrix} \frac{\bar{\mathbf{S}}_{UU}Z}{1-\bar{\mathbf{S}}_{UU}Z} & 0 & 0 \\ 0 & \frac{\bar{\mathbf{S}}_{VV}Z}{1-\bar{\mathbf{S}}_{VV}Z} & 0 \\ 0 & 0 & -1 \end{bmatrix}, \tag{7.18}$$

where all the quantities are computed for $U = X = Y = V = 0$. The surface is locally convex along X when $\bar{\mathbf{S}}_{UU} < 0$ and along Y if $\bar{\mathbf{S}}_{VV} < 0$. The radii of curvature are $R_U(0,0) = \frac{-1}{\bar{\mathbf{S}}_{UU}}$ along X and $R_V = \frac{-1}{\bar{\mathbf{S}}_{VV}}$ along Y. If we substitute also $\|\bar{\mathbf{D}}\| = -Z$ we find that

$$\bar{\mathbf{J}}_{f_\mathsf{D}}(Z) = \begin{bmatrix} \frac{\|\bar{\mathbf{D}}\|}{R_U-\|\bar{\mathbf{D}}\|} & 0 & 0 \\ 0 & \frac{\|\bar{\mathbf{D}}\|}{R_V-\|\bar{\mathbf{D}}\|} & 0 \\ 0 & 0 & -1 \end{bmatrix}. \tag{7.19}$$

Finally the maximum singular value for the Jacobian matrix expressed in the reference frame (X, Y, Z) is

$$\sigma_n(\bar{\mathbf{J}}_{f_\mathsf{D}}(Z)) = |K_\mathrm{V}| \max\left(\frac{\|\bar{\mathbf{D}}\|}{R_U - \|\bar{\mathbf{D}}\|}, \frac{\|\bar{\mathbf{D}}\|}{R_V - \|\bar{\mathbf{D}}\|}, 1\right). \tag{7.20}$$

A basic result of the differential geometry of surfaces can be used to convert this value back to the original reference frame (x, y, z). Since the curvature is an intrinsic property of the surface, the radii of curvature are invariant under rotations, translations, and changes of coordinates. Since $\|\mathbf{D}\| = \|\bar{\mathbf{D}}\|$, no explicit knowledge of the change of coordinate transformation is required to evaluate the Jacobian:

$$\sigma_n(\mathbf{J}_{f_\mathsf{D}}(x, y, z)) = |K_\mathrm{V}| \max\left(\frac{\|\mathbf{D}\|}{r_u(u, v) - \|\mathbf{D}\|}, \frac{\|\mathbf{D}\|}{r_v(u, v) - \|\mathbf{D}\|}, 1\right). \tag{7.21}$$

The values $r_u(u, v)$ and $r_v(u, v)$ are the two principal radii of curvature at the god object point (u, v), with a positive sign assigned for a convex curvature.

A first observation can be made regarding concave or planar surfaces ($r_u, r_v \le 0$): the characteristic number is $q_{f_\mathsf{D}} = \frac{\sigma_n(\mathbf{J}_{f_\mathsf{D}})}{K_\mathrm{V}} = 1$. Therefore, locally, a concave surface has the same passivity properties than a flat virtual wall.

If the point $[u\ v\ \mathbf{S}(u, v)]^\top$ is hyperbolic or convex, one of the radii of curvature is positive. Let's assume, without loss of generality, that $r(u, v)$ is the smallest positive radius of curvature. In this case, $q_{f_\mathsf{D}} = \max(\frac{\|\mathbf{D}\|}{r(u,v)-\|\mathbf{D}\|}, 1)$ is singular for $r(u, v) = \|\mathbf{D}\|$. As the virtual interaction point approaches the center of curvature of the surface $\mathbf{S}$, the norm of the Jacobian increases, because a given movement of the virtual interaction point $\mathbf{x}$ imposes an accelerated motion of the god-object point on the surface $(u, v, \mathbf{S}(u, v))$.

The effect of the curvature becomes significant when $\|\mathbf{D}\| = r(u, v)/2$. To give a specific example, when interacting with a cylinder of radius $r = 10$ mm and stiffness $K_{\mathrm{V}} = 1$ N/mm, a force of 5 N is required to reach the point where $q_{f_{\mathrm{D}}} > 1$.

From this analysis of the characteristic number for a god-object method, it is clear that problems related to non passive behaviors can arise for convex surfaces with high curvature; but the passivity properties are more than acceptable when rendering a larger surface.

7.2.2 Locality of the Characteristic Number

The characteristic number analysis is local, and can be applied to any regular surface $\mathbf{S}$. The results of this analysis need to be interpreted on the basis of the global shape of the surface, which determines the mapping $(x, y, z) \rightarrow (u, v, \mathbf{S}(u, v))$. For example, a convex spherical surface has a singular point at its center, while a convex hemisphere does not. When the virtual interaction point reaches the neighborhood of the center of the hemisphere, the point u, v is on the plane cutting the sphere, and not on the curved portion, which does not corresponds to a singularity.

A hemisphere, however, is not a regular surface, because it has a non-differentiable boundary. Nevertheless, the analysis with the characteristic number is still valid. For concave hemispheres, when the god object point is stuck on the non differentiable boundary the characteristic number is simply $q_{f_{\mathrm{D}}} = 1$ regardless of the radii of curvature at that point.

The latter case shows an example of discontinuities that can arise when using a god-object approach. The characteristic number analysis developed in this chapter is not sensitive to those discontinuous locations. Recall that the medial axis of a surface, $\mathbf{S}$, is the locus of the centers of the maximal bitangent spheres contained in $\mathbf{S}$. When the interaction point is on the medial axis, it is equidistant from at least two locations on the surface, hence when crossing the medial axis, the force changes orientation discontinuously. Not all points of the medial axis, however, are singular points for the characteristic number, neither are all singular points of the characteristic number are on the medial axis. The singular points of the characteristic number are the centers of the tangent spheres of curvatures at the surface $\mathbf{S}$, which are part of the median axis only when the god object point is on a ridge [3].

7.2.3 Effect of Curvature

With the current formalism, the texture map assigns a value of the height field h to a point on the surface $(U, V, \mathbf{S}(U, V))$.

The effect of curvature is expressed by the quantities $U_X = \frac{R_U}{R_U - \|\mathbf{D}\|}$ and $V_Y = \frac{R_V}{R_V - \|\mathbf{D}\|}$, when the virtual interaction point moves inside the surface, $\mathbf{S}$, at a penetration depth, $\|\mathbf{D}\|$, along a curve, $\alpha_{\mathbf{x}}(t) = [x(t)\ y(t)\ z(t)]$. The corresponding trajectory of the god object on the surface is $\alpha_{\mathbf{s}}(t) = [u(t)\ v(t)\ \mathbf{S}(u(t), v(t))]$.

Proposition 1 *If the tangent of the curve $\alpha_{\mathbf{x}}(t)$ is parallel to the tangent plane at $\alpha_{\mathbf{s}}(t)$, then*

$$\|\alpha'_{\mathbf{s}}(t)\| = \left\| \begin{pmatrix} U_X & 0 & 0 \\ 0 & V_X & 0 \\ 0 & 0 & 0 \end{pmatrix} \alpha'_{\mathbf{x}}(t) \right\|. \tag{7.22}$$

Proof An alternative way of describing this effect is to examine the first fundamental form of $\mathbf{S}(u, v)$:

$$I_{uv} = \begin{bmatrix} 1+\mathbf{S}_u^2 & \mathbf{S}_u\mathbf{S}_v \\ \mathbf{S}_u\mathbf{S}_v & 1+\mathbf{S}_v^2 \end{bmatrix}. \tag{7.23}$$

In the normal coordinate frame, the first fundamental form I_{UV} becomes the identity matrix; being $(X, Y) \rightarrow (U, V)$ a change of coordinate for $\mathbf{S}$, the fundamental form becomes $I_{XY} = \left[\begin{smallmatrix} U_X^2 & 0 \\ 0 & V_Y^2 \end{smallmatrix}\right]$, thus showing that lengths in the plane X, Y are distorted on the surface, and this distortion depends on the curvature as well as on the penetration in the surface. □

On a convex surface, in particular, the paths on the surface are longer than the actual motion of the interaction point. The resulting effect is a "compression of the texture" because the ridges are encountered at a higher pace. The opposite effect occurs on concave surfaces.

7.3 Grooved Boundary—Force Normal to Surface (A)—3D Extension

7.3.1 Force Field—3D Flat Plane

The first algorithm under investigation is algorithm **A** whose planar version was already discussed in Sect. 5.5.1 at page 80, [2]. With the planar version on the algorithm, given a planar virtual wall

$$f_{\mathsf{A}}(x, y, z) = K_{\mathrm{V}} \begin{bmatrix} 0 & 0 & -z \end{bmatrix}^{\top} \quad \text{for } z \leq 0. \tag{7.24}$$

The algorithm produces a texture force by replacing the normal penetration in the virtual wall with a penetration, $\mathbf{p} = h(x, y) - z$, normal to a boundary, $h(x, y)$. Then the texture force is

$$f_{\mathsf{A}}(x, y, z) = K_{\mathrm{V}} \begin{bmatrix} 0 \\ 0 \\ h(x, y) - z \end{bmatrix}, \quad \text{for } h(x, y) - z \leq 0. \tag{7.25}$$

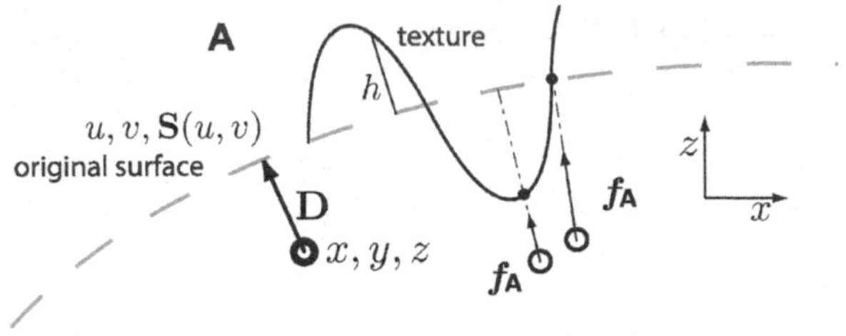

Fig. 7.2 The normal groove algorithm extended to the general 3D surface. The figure shows a x, z slice of the interaction. The force is aligned along the normal to the surface and its magnitude is proportional to the distance with the texture

The Jacobian matrix is easy to compute,

$$\mathbf{J}_{f_\mathbf{A}}(x, y, z) = K_\mathrm{V} \begin{bmatrix} 0 & 0 & 0 \\ 0 & 0 & 0 \\ h_x(x, y) & h_y(x, y) & -1 \end{bmatrix}, \quad \text{for } h(x, y) - z \leq 0 \tag{7.26}$$

and the characteristic number is simply

$$q_{f_\mathbf{A}}(x, y, z) = \sqrt{h_x(x, y)^2 + h_y(x, y)^2 + 1}. \tag{7.27}$$

By extending these results to a generic surface, it will be possible to assess the effect of curvature on the passivity of textures curved surfaces.

7.3.2 Force Field—3D Curved Surface

Figure 7.2 shows the algorithm applied to a generic surface and illustrates the operation of the algorithm. The force always points normally to the surface, and its magnitude changes according to the texture height.

First express the normal to the generic surface **S**,

$$\mathbf{n}(u, v) = \begin{bmatrix} \mathbf{S}_u(u, v) & \mathbf{S}_v(u, v) & -1 \end{bmatrix}^\top. \tag{7.28}$$

Then express the force field,

$$f_\mathbf{A}^\mathbf{S} = K_\mathrm{V} \left[\mathbf{D} + \frac{\mathbf{n}}{\|\mathbf{n}\|} h(j(u, v), k(u, v)) \right]. \tag{7.29}$$

The contact condition for this algorithm becomes $D_\mathbf{n} = \frac{\mathbf{D}^\top \mathbf{n}}{\|\mathbf{n}\|} \geq -h(i, j)$, where $D_\mathbf{n}$ is the normal penetration in the *curved texture surface*.

7.3.3 Change of Coordinates—3D Curved Surface

We now apply the change of coordinates described in the last section to obtain a normal reference frame X, Y, Z [1]. In the new coordinates, the derivatives of the god-object point location with respect to the interaction point can be written

$$U_X(0,0,Z) = \left(1 - \overline{\mathbf{S}}_{UU}(0,0)Z\right)^{-1}, \tag{7.30}$$

$$V_Y(0,0,Z) = \left(1 - \overline{\mathbf{S}}_{VV}(0,0)Z\right)^{-1}. \tag{7.31}$$

The normal has a simplified expression

$$\overline{\mathbf{n}}(0,0,Z) = \begin{bmatrix} 0 & 0 & 1 \end{bmatrix}^{\mathsf{T}}, \tag{7.32}$$

$$\begin{aligned}
\overline{\mathbf{J}}_{\mathbf{n}}(0,0,Z) = \overline{\mathbf{n}}_{\{X,Y,Z\}}(0,0,Z) &= -\begin{bmatrix} \overline{\mathbf{S}}_{UX} & \overline{\mathbf{S}}_{UY} & \overline{\mathbf{S}}_{UZ} \\ \overline{\mathbf{S}}_{VX} & \overline{\mathbf{S}}_{VY} & \overline{\mathbf{S}}_{VZ} \\ 0 & 0 & 0 \end{bmatrix} \\
&= -\begin{bmatrix} \frac{\overline{\mathbf{S}}_{UU}(0,0)}{1-\overline{\mathbf{S}}_{UU}(0,0)Z} & 0 & 0 \\ 0 & \frac{\overline{\mathbf{S}}_{VV}(0,0)}{1-\overline{\mathbf{S}}_{VV}(0,0)Z} & 0 \\ 0 & 0 & 0 \end{bmatrix}
\end{aligned} \tag{7.33}$$

and the norm of the normal vector adopts a trivial form

$$\|\overline{\mathbf{n}}(0,0,Z)\| = \left(1 + \overline{\mathbf{S}}_U(0,0)^2 + \overline{\mathbf{S}}_V(0,0)^2\right)^{1/2} = 1, \tag{7.34}$$

$$\begin{aligned}
\|\overline{\mathbf{n}}(0,0,Z)\|_{\{X,Y,Z\}} &= \frac{\begin{bmatrix} \overline{\mathbf{S}}_U(0,0)\overline{\mathbf{S}}_{UX}(0,0) + \overline{\mathbf{S}}_V(0,0)\overline{\mathbf{S}}_{VX}(0,0) \\ \overline{\mathbf{S}}_U(0,0)\overline{\mathbf{S}}_{UY}(0,0) + \overline{\mathbf{S}}_V(0,0)\overline{\mathbf{S}}_{VY}(0,0) \\ \overline{\mathbf{S}}_U(0,0)\overline{\mathbf{S}}_{UZ}(0,0) + \overline{\mathbf{S}}_V(0,0)\overline{\mathbf{S}}_{VZ}(0,0) \end{bmatrix}}{\|\overline{\mathbf{n}}\|} \\
&= \begin{bmatrix} 0 & 0 & 0 \end{bmatrix}^{\mathsf{T}}.
\end{aligned} \tag{7.35}$$

Finally it is possible to write $D_{\mathbf{n}} = -Z$, which simplifies the specification of the contact condition. A point is inside the texture if $Z \le h(i, j)$.

7.3.4 *Jacobian—3D Curved Surface*

To compute the Jacobian matrix, first apply the chain rule,

$$\overline{\mathbf{J}}_{f_{\mathsf{A}}}^{\mathbf{S}} = \overline{\mathbf{J}}_{f_{\mathsf{D}}} + K_{\mathrm{V}}\left[\overline{\mathbf{n}}_{\{X,Y,Z\}} h(i(U,V), j(U,V)) + \overline{\mathbf{n}}_{\{X,Y,Z\}} h(i(U,V), j(U,V))\right]. \tag{7.36}$$

The previous simplifications give

$$\bar{\mathbf{J}}^{\mathbf{S}}_{f_{\mathbf{A}}} = \bar{\mathbf{J}}_{f_{\mathbf{D}}} + K_{\mathrm{V}} \begin{bmatrix} -\frac{\bar{\mathbf{S}}_{UU}(0,0)h(0,0)}{1-\bar{\mathbf{S}}_{UU}(0,0)Z} & 0 & 0 \\ 0 & -\frac{\bar{\mathbf{S}}_{VV}(0,0)h(0,0)}{1-\bar{\mathbf{S}}_{VV}(0,0)Z} & 0 \\ h_X(0,0) & h_Y(0,0) & 0 \end{bmatrix} \tag{7.37}$$

$$= \bar{\mathbf{J}}_{f_{\mathbf{D}}} + K_{\mathrm{V}} \begin{bmatrix} \frac{\bar{\mathbf{S}}_{UU}(-h)}{1-\bar{\mathbf{S}}_{UU}Z} & 0 & 0 \\ 0 & \frac{\bar{\mathbf{S}}_{VV}(-h)}{1-\bar{\mathbf{S}}_{VV}Z} & 0 \\ (h_i i_U + h_j j_U)U_X & (h_i i_V + h_j j_V)V_Y & 0 \end{bmatrix} \tag{7.38}$$

$$= \bar{\mathbf{J}}_{f_{\mathbf{D}}} + K_{\mathrm{V}} \begin{bmatrix} \frac{\bar{\mathbf{S}}_{UU}(-h)}{1-\bar{\mathbf{S}}_{UU}Z} & 0 & 0 \\ 0 & \frac{\bar{\mathbf{S}}_{VV}(-h)}{1-\bar{\mathbf{S}}_{VV}Z} & 0 \\ \frac{h_i i_U + h_j j_U}{1-\bar{\mathbf{S}}_{UU}Z} & \frac{h_i i_V + h_j j_V}{1-\bar{\mathbf{S}}_{VV}Z} & 0 \end{bmatrix} \tag{7.39}$$

$$= K_{\mathrm{V}} \begin{bmatrix} \frac{-D_{\mathbf{n}}-h}{R_U-D_{\mathbf{n}}} & 0 & 0 \\ 0 & \frac{-D_{\mathbf{n}}-h}{R_V-D_{\mathbf{n}}} & 0 \\ \frac{R_U(h_i i_U + h_j j_U)}{R_U-D_{\mathbf{n}}} & \frac{R_V(h_i i_V + h_j j_V)}{R_V-D_{\mathbf{n}}} & -1 \end{bmatrix} \tag{7.40}$$

$$= K_{\mathrm{V}} \begin{bmatrix} \frac{-D_{\mathbf{n}}-h}{R_U-D_{\mathbf{n}}} & 0 & 0 \\ 0 & \frac{-D_{\mathbf{n}}-h}{R_V-D_{\mathbf{n}}} & 0 \\ 0 & 0 & 0 \end{bmatrix} + K_{\mathrm{V}} \begin{bmatrix} 0 & 0 & 0 \\ 0 & 0 & 0 \\ \frac{R_U(h_i i_U + h_j j_U)}{R_U-D_{\mathbf{n}}} & \frac{R_V(h_i i_V + h_j j_V)}{R_V-D_{\mathbf{n}}} & -1 \end{bmatrix},$$

where R_U and R_V are the two principal radii of curvature. The matrix $\mathbf{M} = \left[\begin{smallmatrix} i_U & i_V \\ j_U & j_V \end{smallmatrix}\right]$ is the Jacobian of the local texture map which assigns a point on the surface $\mathbf{S}$ (parametrized along the principal directions) to a point on the plane where $h(i, j)$ is defined. The properties of this map are very interesting and are reflected in the structure of the matrix $\mathbf{M}$. If the map is locally conformal at the point (u, v) then $\mathbf{M}^\top \mathbf{M} = \left[\begin{smallmatrix} t^2 & 0 \\ 0 & t^2 \end{smallmatrix}\right]$, with $t = 1$ for a locally isometric map.

To simplify the analysis, the Jacobian can be separated into two matrices. The characteristic number is only an approximation but can provide many insights. For instance, it is possible to find an upper bound

$$\max\left(\frac{\|-D_{\mathbf{n}}-h\|}{\|r_u - D_{\mathbf{n}}\|}, \frac{\|-D_{\mathbf{n}}-h\|}{\|r_v - D_{\mathbf{n}}\|}\right) + \sqrt{\left(\frac{r_v(h_i i_V + h_j j_V)}{r_v - D_{\mathbf{n}}}\right)^2 + \left(\frac{r_u(h_i i_U + h_j j_U)}{r_u - D_{\mathbf{n}}}\right)^2 + 1} \tag{7.41}$$

and a lower bound

$$\max\left(\sqrt{\left(\frac{r_u(h_i i_U + h_j j_U)}{r_u - D_\mathbf{n}}\right)^2 + \left(\frac{-D_\mathbf{n} - h}{r_u - D_\mathbf{n}}\right)^2},\right.$$
$$\left.\sqrt{\left(\frac{r_v(h_i i_V + h_j j_V)}{r_v - D_\mathbf{n}}\right)^2 + \left(\frac{-D_\mathbf{n} - h}{r_v - D_\mathbf{n}}\right)^2}, 1\right). \tag{7.42}$$

Because the penetration in the untextured surface $\mathbf{S}$ can be negative $D_\mathbf{n} \geq -h(i,j)$, the added texture could lead to a singularity also for a concave surface when $D_\mathbf{n} = R(u,v)$ with $R \leq 0$. In particular it would happen if $h(i,j) \leq R(u,v)$. This situation is unlikely because it would require the application of a texture as large as the underlying surface, which could generate artifacts such as the interpenetration of two parts of the texture. In general we assume $|h(i,j)| \ll |R(u,v)|$.

7.3.4.1 Cylinder with Sinusoidal Texture

For the sake of example, consider the case of a cylindrical surface $\mathbf{S}$ with radii $r_u \neq \infty, r_v \to \infty$ textured with a sinusoidal height field $h(i,j) = A\sin(2\pi i/L + \phi)$ of amplitude A, spatial period L and phase ϕ. Since the cylinder is developable, the mapping of a planar texture over the cylinder is globally isometric, hence, we can set $\mathbf{M} = \left[\begin{smallmatrix}1 & 0\\ 0 & 1\end{smallmatrix}\right]$, with the effect of making the ridges of the texture aligned with the axis of the cylinder. The Jacobian for a cylinder and isometric texture becomes

$$\mathbf{J}_{f_\mathbf{A}}^{\mathbf{S}} = K_\mathrm{V}\begin{bmatrix} \frac{-D_\mathbf{n}-h}{r_u-D_\mathbf{n}} & 0 & 0 \\ 0 & 0 & 0 \\ \frac{r_u h_i}{r_u-D_\mathbf{n}} & h_j & -1 \end{bmatrix}. \tag{7.43}$$

As expected, when $r_u \to \infty$, the characteristic number simplifies to (7.27). For the unidirectional sinusoidal texture

$$q_{f_\mathbf{A}}^{\mathbf{S}} \geq \max\left(\sqrt{\left(\frac{r_u 2\pi A\cos(\theta+\phi)}{L(r_u - D_\mathbf{n})}\right)^2 + \left(\frac{-D_\mathbf{n} - A\sin(\theta+\phi)}{r_u - D_\mathbf{n}}\right)^2}, 1\right),$$

$$q_{f_\mathbf{A}}^{\mathbf{S}} \leq \left\|\frac{-D_\mathbf{n} - A\sin(\theta+\phi)}{r_u - D_\mathbf{n}}\right\| + \left(\sqrt{\left(\frac{r_u 2\pi A\cos(\theta+\phi)}{L(r_u - D_\mathbf{n})}\right)^2 + 1}\right), \tag{7.44}$$

where θ is the phase of the sine wave at the point u, v on the surface. This result is consistent with the planar case since when $r_u \to \infty$, the characteristic number simplifies to

$$\max\left(\sqrt{\left(\frac{2\pi A\cos(\theta+\phi)}{L}\right)^2}, 1\right) \leq q_{f_\mathbf{A}}^{\mathbf{S}} \leq \sqrt{\left(\frac{2\pi A\cos(\theta+\phi)}{L}\right)^2 + 1}. \tag{7.45}$$

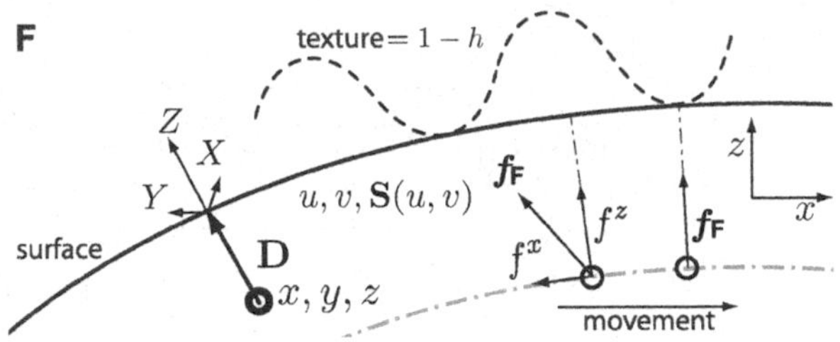

Fig. 7.3 Extensions of the friction algorithm to a generic curved surface. The figure shows a x, z slice of the haptic interaction. The normal penalty force is computed from the boundary of the surface **S**, while the lateral force is modulated according to the texture height field

It is interesting to notice that if the texture has an extremely large spatial period, $L \to \infty$, the almost flat texture still modifies the characteristic number to become

$$q_{f_\mathsf{A}}^{\mathbf{S}} = \max\left(\left\| \frac{-D_\mathbf{n} - A\sin(+\phi)}{r_u D_\mathbf{n}} \right\|, 1\right). \tag{7.46}$$

In addition, a texture applied to a cylinder does not change the singular points, they remain located on the axis of the untextured cylinder.

7.4 Friction Algorithm—Extension to 3D

The friction texture algorithm introduced in Sect. 5.5.6 at page 86 can be extended to curved surfaces, but this generalization is not as straightforward as it is for the normal penetration algorithm just discussed. The main idea is to generate a friction field tangential to the surface **S** and then modulate it based on the height field of the texture, see Fig. 7.3. The first step is to extend the friction generation algorithm to a general curved surface.

7.4.1 Friction Field

The friction algorithm described by Hayward and Armstrong is modeled as an hysteretic elastic contact between the virtual interaction point and an anchor point on a virtual planar surface [4]. When the pre-sliding displacement of this contact reaches a breakaway threshold, the anchor point is dragged on the plane. In general, the anchor point is displaced only when the distance between the virtual interaction point and the anchor point is greater than a distance $d_{\max}$.

The extension of this elastic contact model on the surface requires the computation of the *geodesic distance* between the anchor point, $\mathbf{w_S}$, and the interaction point, $\mathbf{x_S}$, both of which are on the surface **S**. When this distance exceeds $d_{\max}$, the anchor point is moved *along the geodesic curve* joining $\mathbf{w_S}$ and $\mathbf{x_S}$ so that the new anchor $\|\mathbf{w}_\mathbf{S}^{\mathrm{new}} - \mathbf{x_S}\|_{\mathrm{geo}} = d_{\max}$.

If $\alpha(t)$, $0 \le t \le 1$, is the geodesic curve such that $\alpha(0) = \mathbf{w_S}$ and $\alpha(1) = \mathbf{x_S}$ then it is possible to define the planar equivalent of the geodesic to be

$$\mathbf{D}_\alpha = \frac{\alpha'(1)}{\|\alpha'(1)\|} \int_0^1 |\alpha'(\tau)| \mathrm{d}\tau, \tag{7.47}$$

where $|\mathbf{D}_\alpha| \le d_{\max}$. When the threshold is exceeded, then $\mathbf{w}_\mathbf{S}^{\text{new}} = \alpha(t)$ with t such that $\int_t^1 |\alpha'(\tau)| \mathrm{d}\tau = d_{\max}$. In the case of the god-object, given a regular surface $\mathbf{S}$, the interaction point $\mathbf{x}$, and the god object point $\mathbf{s}$, the deflection is

$$\mathbf{D} = \mathbf{D}(\mathbf{x}) = \mathbf{s} - \mathbf{x} = \begin{bmatrix} u - x \\ v - y \\ \mathbf{S}(u, v) - z \end{bmatrix}, \tag{7.48}$$

that is, the distance between the interaction point and the surface. A basis of the tangent space at point $\mathbf{s}$ is $(\mathbf{T}_u, \mathbf{T}_v)$, where

$$\mathbf{T}_u = \begin{bmatrix} 1 \\ 0 \\ S_u \end{bmatrix} \quad \text{and} \quad \mathbf{T}_v = \begin{bmatrix} 0 \\ 1 \\ S_v \end{bmatrix}. \tag{7.49}$$

We can then express the geodesic distance in the tangent plane

$$\mathbf{D}_\alpha = \frac{[a\mathbf{T}_u + b\mathbf{T}_v]}{\|\mathbf{T}_u\| \|\mathbf{T}_v\|}. \tag{7.50}$$

The friction force is tangent to the curve and should be proportional to the magnitude of normal force $f_\mathbf{D}$. According to this constraints

$$f_\mathrm{f} = -\frac{\mu K_\mathrm{V} \|\mathbf{D}\|}{d_{\max}} \frac{[a\mathbf{T}_u + b\mathbf{T}_v]}{\|\mathbf{T}_u\| \|\mathbf{T}_v\|}. \tag{7.51}$$

7.4.2 Jacobian of the Friction Field

In the normal coordinate frame (X, Y, Z) the following properties hold:

$$\|\mathbf{T}_U\| = 1, \tag{7.52}$$

$$\|\mathbf{T}_V\| = 1, \tag{7.53}$$

$$\|\mathbf{T}_u\|_{\{X,Y,Z\}} = \|\mathbf{T}_v\|_{\{X,Y,Z\}} = \begin{bmatrix} 0 & 0 & 0 \end{bmatrix}^\mathrm{T}, \tag{7.54}$$

$$\mathbf{T}_u = \begin{bmatrix} 1 & 0 & 0 \end{bmatrix}^\mathrm{T}, \qquad \mathbf{T}_v = \begin{bmatrix} 0 & 1 & 0 \end{bmatrix}^\mathrm{T}, \tag{7.55}$$

$$(a\mathbf{T}_u + b\mathbf{T}_v)_{\{X,Y,Z\}} = \begin{bmatrix} a_U U_X & a_V V_Y & 0 \\ b_U U_X & b_V V_Y & 0 \\ a_U \overline{\mathbf{S}}_{UX} & b_V \overline{\mathbf{S}}_{VX} & 0 \end{bmatrix} \tag{7.56}$$

$$(a\mathbf{T}_u + b\mathbf{T}_v)_{\{X,Y,Z\}} = \begin{bmatrix} a_U U_X & 0 & 0 \\ 0 & b_V V_Y & 0 \\ a_U \overline{\mathbf{S}}_{UU} U_X & b_V \overline{\mathbf{S}}_{VV} V_Y & 0 \end{bmatrix}, \tag{7.57}$$

where the change in geodesic length has been approximated by a planar friction field, which is a linear force-displacement relationship, thus $b_U = a_V = 0$.

This approximation is valid only if the elastic threshold $d_{\max}$ is small compared to the curvature of the surface **S**, and would fail otherwise. For example, on a sphere with radius $r = d_{\max}/\pi$ the interaction point reaches the antipode of the anchor point at maximum deflection. There, a small variation in the position of the interaction point would result in a large change in the orientation of the force, which is not captured by the simplification above.

In general, this scenario is highly unlikely because $d_{\max}$ should be very small, in the order of 500 μm. To avoid artifacts, the surfaces should have a radius in the order of 10 mm or more. Moreover, for a spherical surface of radius R and for $d_{\max} \le R/20$ the difference between the Euclidean distance and the geodesic distance is small, $\|\mathbf{w}_{\mathbf{S}}^{\text{new}} - \mathbf{x}_{\mathbf{S}}\|_{\text{geo}} \le d_{\max} \le R/20$ and the length of the segment conjoining the two point is $\|\mathbf{w}_{\mathbf{S}}^{\text{new}} - \mathbf{x}_{\mathbf{S}}\| \le R\sin(1/20) \approx R(1/20)$. Finally, the curvature restriction over the supporting surface does not pertain the height field h.

The Jacobian matrix of the force field $\boldsymbol{f} = K_{\text{V}}\overline{\mathbf{D}} + \boldsymbol{f}_{\text{f}}$ can be finally expressed as

$$\overline{\mathbf{J}}_{f_{\mathsf{F}}}^{\mathbf{S}} = \overline{\mathbf{J}}_{f_{\mathsf{D}}} - \frac{\mu K_{\text{V}}\|\mathbf{D}\|}{d_{\max}} \begin{bmatrix} a_U U_X & 0 & 0 \\ 0 & b_V V_Y & 0 \\ a_U \overline{\mathbf{S}}_{UU} U_X & b_V \overline{\mathbf{S}}_{VV} V_Y & 0 \end{bmatrix} - \frac{\mu K_{\text{V}}}{d_{\max}} \begin{bmatrix} 0 & 0 & a \\ 0 & 0 & b \\ 0 & 0 & 0 \end{bmatrix}. \tag{7.58}$$

As for the unidirectional planar case, the friction algorithm imposes constraints on the elastic deflection. If the contact is not at presliding threshold $\sqrt{a^2+b^2} < d_{\max}$ and $a_U = b_V = 1$, during the sliding condition $a_U = b_V = 0$ and $\sqrt{a^2+b^2} = d_{\max}$.

7.4.3 Example: Cylinder with Friction

The force field for a cylinder of radius $|R_U| < \infty$, with unidirectional friction around the cylinder and no friction along it ($b_V = 0$ and $b = 0$) has the following Jacobian matrix,

$$\overline{\mathbf{J}}_{f_{\mathsf{F}}}^{\mathbf{S}} = K_{\text{V}} \begin{bmatrix} \frac{1-\mu a_U R_U/d_{\max}}{R_U - \|\mathbf{D}\|}\|\mathbf{D}\| & 0 & \frac{\mu a}{d_{\max}} \\ 0 & 0 & 0 \\ \frac{-\mu a_U}{d_{\max}(R_U - \|\mathbf{D}\|)}\|\mathbf{D}\| & 0 & -1 \end{bmatrix}. \tag{7.59}$$

Despite the complexity of the matrix, it is possible to identify the a problematic behavior similar to that of the unidirectional flat case explored in Chap. 5. The term

$\frac{\mu a_U R_U/d_{max}}{R_U-\|\mathbf{D}\|}\|\mathbf{D}\|$ dominates the Jacobian for any value of R_U during the stuck phase of the friction algorithm ($a_u = 1$) and it grows as the penetration depth does. Clamping the value of $\|\mathbf{D}\|$ is mandatory both in the convex and concave case.

7.5 Modulated Lateral Friction (F)—3D Extension

The texture sensation is created by modulating the friction force with the texture height field, with the further condition $|h(i, j)| \leq 1$. The resulting force field is

$$f_{\mathsf{F}} = f_{\mathsf{D}} + (1 - h(i, j)) f_{\mathsf{f}}. \tag{7.60}$$

It has a fairly involved Jacobian matrix

$$\bar{\mathbf{J}}_{f_{\mathsf{F}}}^{\mathbf{S}} = \bar{\mathbf{J}}_{f_{\mathsf{D}}} - \frac{(1 - h(i, j))\mu K_{\mathrm{V}}}{d_{max}} \begin{bmatrix} a_U U_X\|\mathbf{D}\| & 0 & a \\ 0 & b_V V_Y\|\mathbf{D}\| & b \\ a_U \bar{\mathbf{S}}_{UU} U_X\|\mathbf{D}\| & b_V \bar{\mathbf{S}}_{VV} V_Y\|\mathbf{D}\| & 0 \end{bmatrix}$$
$$+ \frac{\mu K_{\mathrm{V}}\|\mathbf{D}\|}{d_{max}} \begin{bmatrix} a(h_i i_U + h_j j_U)U_X & a(h_i i_V + h_j j_V)V_Y & 0 \\ b(h_i i_U + h_j j_U)U_X & b(h_i i_V + h_j j_V)V_Y & 0 \\ 0 & 0 & 0 \end{bmatrix}, \tag{7.61}$$

where the elements of the matrix $\mathbf{M} = \left[\begin{smallmatrix} i_U & i_V \\ j_U & j_V \end{smallmatrix}\right]$ describe the structure of the local texture map between the surface $\mathbf{S}$ and the planar height field.

7.5.1 Example: Cylinder with Friction, Sinusoidal Texture

The friction-based texture algorithm is applied to the same cylindrical surface $\mathbf{S}$ with radii $R_U \neq \infty$, $R_V = \infty$ textured with a sinusoidal height field $h(i, j) = A\sin(2\pi i/L)$ of amplitude A, and spatial period L. The surface is developable, hence the mapping with the plane is globally isometric, leading to a simple texture map $\mathbf{M} = \left[\begin{smallmatrix} 1 & 0 \\ 0 & 1 \end{smallmatrix}\right]$. The friction is unidirectional, with no component along the axis of the cylinder ($b = b_V = 0$), and the Jacobian becomes

$$\bar{\mathbf{J}}_{f_{\mathsf{F}}}^{\mathbf{S}} = K_{\mathrm{V}} \begin{bmatrix} \frac{1+[ah_i - a_U + ha_U]\mu R_U/d_{max}]}{R_U - \|\mathbf{D}\|}\|\mathbf{D}\| & 0 & (1-h)\frac{\mu a}{d_{max}} \\ 0 & 0 & 0 \\ \frac{-(1-h)\mu a_U}{d_{max}(R_U - \|\mathbf{D}\|)}\|\mathbf{D}\| & 0 & -1 \end{bmatrix} \tag{7.62}$$

which is again dominated by a term dependent on the penetration in the virtual surface.

7.6 Conclusion

In this chapter, the characteristic number analysis is extended to generic 3D textures, with the goal of assessing the role of curvature over the passivity margins of haptic textures. The first notable result is the presence of singularity points inside convex haptic surfaces which are simulated with a standard, penalty based, god-object style algorithm. On the other hand, the haptic interaction with flat or concave surfaces does not reduce passivity. Apart from the singularities of convex surfaces, the unidirectional-flat plane analysis carried out in the previous chapter identified the major sources of non passivity.

A notable effect of curvature is the compression of the textures: when exploring a superficial texture with spatial period L with a velocity v, the user experiences a textural signal with a frequency that is distorted. On convex surfaces the frequency is increased, on concave surfaces the frequency is decreased.

References

1. Carmo, M.D.: Differential Geometry of Curves and Surfaces. Prentice Hall Inc., Englewood Cliffs (1976)
2. Choi, S., Tan, H.Z.: Perceived instability of virtual haptic texture. I. Experimental studies. Presence **13**(4), 395–415 (2004)
3. Giblin, P., Kimia, B.B.: A formal classification of 3d medial axis points and their local geometry. IEEE Trans. Pattern Anal. Mach. Intell. **26**(2), 238–251 (2004)
4. Hayward, V., Armstrong, B.: A new computational model of friction applied to haptic rendering. In: Corke, P., Trevelyan, J. (eds.) Experimental Robotics VI. Lecture Notes in Control and Information Sciences, vol. 250, pp. 403–412 (2000)
5. Hearn, D., Baker, M.P.: Computer Graphics (c Version). Prentice Hall, New York (1996)
6. Zilles, C.B., Salisbury, J.K.: A constraint-based god object method for haptic display. In: Proceedings of the IEEE/RSJ International Conference on Intelligent Robots and Systems, IROS'95, vol. 3, pp. 146–151 (1995)

Chapter 8
Roughness of Virtual Textures and Lateral Force Modulation

Abstract We describe experiments that compared the perceived relative roughness of textured virtual walls synthesized with an accurately controlled haptic interface. Texture was modeled as a spatially modulated sinusoidal friction grating. The results indicate that both the modulation depth of the grating (A), and the coefficient of friction (μ) are strongly associated with the perceived roughness when increasing either A or μ. Changing the spatial period of the grating (l), however, did not yield consistent relative roughness judgement results, indicating that there is a weaker association.

8.1 Preface to Chap. 8

The theoretical and experimental work presented in the previous chapters allows the rendering of precise haptic textures in terms of frequency and stability regardless of the algorithm used to generate the texture. This quantitative validation, however, does not explain the perceptual response of the user to textures generated with any given algorithms; the perceptual characteristics of algorithms must then be investigated with a psychphysic experiment. The percept of roughness is central to the perception of the superficial properties of objects, hence it is important to understand the relationship between roughness and the texturing algorithm used.

This chapter introduces a fundamental property of the friction based texture rendering algorithm. The perception of roughness scales monotonically with two of the parameters describing the textural elements: the amplitude of the modulation of the texture and the friction coefficient of the surface. The pitch did not exhibit a consistent trend, in accord with previous studies.

These results were essential in order to simplify the design of the roughness calibration experiment presented in the next chapter, because a monotonic relationship between parameters and percepts allows the use of fast adaptive staircase thresholding methods.

Reprinted from Gianni Campion, Andrew H.C. Gosline, and Vincent Hayward, "Does Judgement of Haptic Virtual Texture Roughness Scale Monotonically with Lateral Force Modulation?" *Lecture Notes in Computer Science*, Volume 5024/2008, Pages 718–723, 2008.

G. Campion, *The Synthesis of Three Dimensional Haptic Textures: Geometry, Control, and Psychophysics*, Springer Series on Touch and Haptic Systems,
DOI 10.1007/978-0-85729-576-7_8,

8.1.1 Contribution of Authors

Gianni Campion designed, implemented, and ran the experiment; he also analyzed the data and prepared the figures, and wrote the manuscript. Andrew Gosline designed and fabricated the two non programmable dampers and the blades used to stabilize the haptic rendering and edited the manuscript. While not essential for the haptic rendering of textures, the psychophysical experiment benefited from the addition of the dampers because they increased the range of impedances of the Pantograph. Prof. Hayward supervised and commented the work, edited the manuscript and the figures.

8.2 Introduction

Much work was reported regarding virtual haptic textures and the experience of roughness that results from a variety of synthesis approaches. To our knowledge, however, no previous study addresses the question of whether the experience of roughness scales monotonically with synthesis parameters. The property of perceptual monotonicity could greatly simplify the search space for the investigation of equivalent sensations given by different hardware/software combinations. To this end, we use a friction-based algorithm that provides a sensation of texture. Mechanical signals are delivered a high-quality device that is engineered so that its dynamic characteristics are unlikely to interfere with the subjective results. Actuation and sensing resolution of the device allow for a precise reproduction of the texture, and passivity theory is applied to ensure the quality of the synthetic texture.

The perception of the roughness of virtual haptic textures has been extensively explored but the results are difficult to compare. Early work by Lederman and Minsky showed that the roughness estimate of 2D synthetic virtual textures could be almost entirely predicted by the maximum lateral force [10]. A recent study by Kornbrot et al. explored the psychometric function linking perceived roughness and the spatial frequency of virtual sinusoidal textures [8]; in their results, the majority of subjects had a descending function (larger pitch correlated with smaller roughness), while other studies indicated either raising or quadratic relationships. A descending trend has been reported, for example by Wall et al. [12], while for physical textures a quadratic function was found [7]. A further complication arises in the very definition of "virtual roughness". This difficulty is apparent in such works as [9], where the authors felt it necessary to clarify what roughness was by comparing the haptic experience with that of a car running on a bumpy road.

The goal of the present study is to validate the authors' observation that friction-based textured surfaces, synthesized according to a force field of the form (8.1), feel rougher when either the depth of modulation, A, or the coefficient of friction, μ, increases monotonically.

$$F_{\text{lateral}} = \mu F_{\text{normal}}(1 - A\sin(2\pi x/l)). \tag{8.1}$$

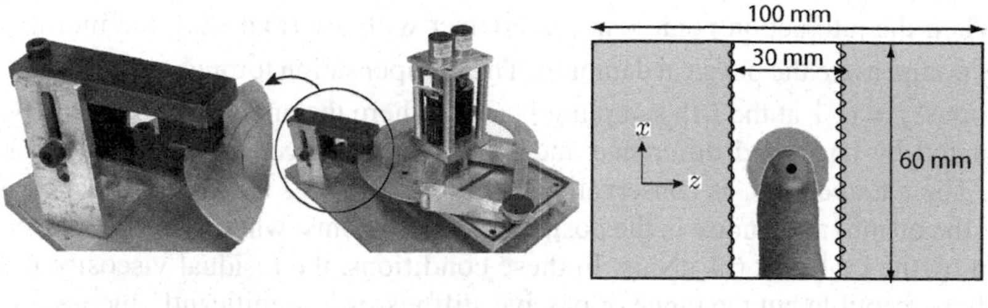

Fig. 8.1 *Left:* A direct-driven five-bar mechanism drives a small plate in the horizontal plane. In this version of the Pantograph, each joint has a eddy-current viscous damper. *Right:* Subjects were presented with two textured walls and asked which one was rougher. The virtual interaction point is marked by a *black circle*, and the free space is in *white*

To reach this objective, a minimalist definition of roughness was given to the subjects, "roughness is the opposite of smoothness". This definition was thought to minimize subjective bias, yet it prevented confusion. This intuition is consistent with previous findings by Smith et al. since the rate-of-change of the textural force is directly correlated with both quantities [11]. Since the rate of change has the form $\propto A\mu/l$, the effects of the spatial frequency are also explored.

8.3 Device and Control

We used a Pantograph haptic device, further described in [2]. It can display forces up to 2 N with a spatial resolution of 0.01 mm in a horizontal workspace of 100×60 mm. It has flat structural response within DC–400 Hz. Simulating dry friction is equivalent to creating high-gain force-feedback [5]. In a sampled-data setting, to guarantee the synthesis to be passive, this gain is limited by the physical dissipation of the device [3]. Our basic device has very little of it, limiting the range of parameters that can be tested. This problem is compounded by the use of 400 Hz cut-off reconstruction filter that adds additional phase lag [1].

We retrofitted the device with eddy-current brakes that add accurately defined viscosity [4]. The magnetic field is produced by C-clamp magnetic circuits terminating with a pair of 11 mm cubic-shaped rare earth magnets (NdFeB, Amazing Magnets LLC, Irvine, CA, USA), as shown in Fig. 8.1, left panel. Blades, in the shape of annulus arcs, are affixed to the proximal arms of the device and move in the air gap, creating a viscous torque that can be adjusted by tuning the amount of overlap between the gap and the blade. This way, the "base dynamics" of the device feel like a nonuniform viscous field.

To compensate for this viscous field in free space, we adopted an approach suggested by Colgate and Shenkel [3]. Assistive virtual damping was used to compensate the extra physical damping; since the sampled data approximation always errs on the side of adding energy, its negation errs on the side of not removing enough of it, thus preserving passivity. We applied this approach to our hardware. To our knowledge, it is the first report of its successful implementation.

When the interaction point is not in contact with a virtual wall, the motors partially compensate the physical damping. The compensation torques are $\tau_i^j = +B_c \omega_i^j$ for joints $j = 0, 1$ at the i-th sampling period, where the angular velocities, ω^j, are estimated by backward difference and then averaged over a window of 24 samples. The parameter B_c is conservatively set to 4.5 mNms to avoid artifacts arising from the quantization noise in the position measurements, where the actual viscosity given by the brakes is 6.1 nNms. In these conditions, the residual viscosity field is barely perceptible but the range of passive stiffnesses is significantly increased.

8.4 Texture Force Field

The Pantograph device can render planar forces in a z, x plane; thus, the user experiences 1D constraints with 1D textures, see Fig. 8.1 for the axes definitions. First, a virtual wall is synthesized for $z \le z_{\text{wall}}$

$$F_z = -K d^z, \quad \text{if } z \le z_{\text{wall}}; \qquad F_z = 0, \quad \text{otherwise}, \tag{8.2}$$

where $d^z = z - z_{\text{wall}}$ is the (negative) penetration in the virtual wall and K the stiffness. Then, a time-free, that is velocity-independent, dry-friction algorithm is used to compute a static friction field with a coefficient μ [6]:

$$F_{\text{friction}} = \underbrace{-\mu\, K \underbrace{\max(d^z, d^z_{\max})}_{\text{penetration clamp}}}_{\text{normal force}} \underbrace{\frac{\min(d^x, d^x_{\max})}{d^x_{\max}}}_{\text{friction algorithm}}, \tag{8.3}$$

where: $d^x \le d^x_{\max}$ is the pre-sliding tangential deflection and $d^z_{\max}$ is used to ensure passive synthesis by limiting the maximum friction force. Finally, this friction force field is modulated with a sinusoidal generating function:

$$F_x = F_{\text{textured}} = F_{\text{friction}}(1 - A \sin(2\pi x / l)), \tag{8.4}$$

where l is the spatial frequency of the texture and $0 \le A \le 1$ is the modulation depth of the textural force. The same algorithm is used on both sides of the workspace, $z \ge z_{\text{wall}}$ and $z \le -z_{\text{wall}}$.

The values $K = 1$ Nmm, $d^x_{\max} = d^z_{\max} = 0.5$ mm were selected because no limit cycles were present when exploring the surfaces, and because they offered a good trade-off between fidelity of the friction model and maximum lateral force.

8.5 Experimental Procedure

8.5.1 Design

To test our hypothesis, the influence of A, μ, and l is tested independently. Two 1-D virtual walls, facing each other and spaced 30 mm ($z_{\text{wall}} = \pm 15$ mm), were available to the subjects, Fig. 8.1, right panel. Each had a sinusoidal texture, and subjects were to identify which of the two surfaces felt rougher: left or right? The two textures differed by the value of exactly one parameter, either A, μ, or l. The answer was entered by a keystroke and the time of each trial was recorded. The two textures were always different.

8.5.2 Stimuli

Three sets of stimuli were prepared, one for each parameter. Six pairs of textures were presented for amplitude $A \in \{0.25, 0.5, 0.75, 1\}$: $(A_1, A_2) \in \{(1, 0.75), (1, 0.5), (1, 0.25), (0.75, 0.5), (0.75, .25), (0.5, 0.25)\}$. These six amplitude pairs were tested in nine different conditions: $l \in \{1, 2, 3\} \times \mu \in \{1/3, 2/3, 1\}$ for a total of 54 trials for these parameters. With the same procedure, 54 trials were prepared for testing the parameter $l \in \{1, 2, 3, 4\}$ mm in the nine conditions $\mu \in \{1/3, 2/3, 1\} \times A \in \{1/3, 2/3, 1\}$; also the combinations of $\mu \in \{1, 2, 3, 4\}$ were presented in the nine conditions $l \in \{1, 2, 3\} \times A \in \{1/3, 2/3, 1\}$. Each subject performed 162 pairwise comparisons; the surface textures were randomly assigned to the left or right wall.

8.5.3 Subjects

Six right handed subjects, two female and four male, volunteered for the experiment; among them were two authors of this paper. Four of the subjects were very familiar with haptic technology. No definition of roughness was given to the subjects except that "roughness was the opposite of smoothness". There was no training and no feedback was provided during the tests. Subjects wore sound isolation headphones (DirectSound EX-29), and white noise was played to mask the sound of the device. Most subjects reported ambiguity when dealing with textures with different frequencies.

8.6 Results

Figure 8.2 shows the percentage of times each subject responded in agreement with the hypothesis of monotonicity. In total, the hypothesis was confirmed in almost 97% of the 324 trials for parameter A: only 10 times the surface with a smaller A was identified as rougher. Similar results hold for μ: 9 disagreements over 324 trials. The previous literature indicates that roughness decreases with the spatial period. The experiment confirms it, but the agreement drops to 80%. Equivalently, 20% of times the subject chose the spatial period to be rougher. This effect could be

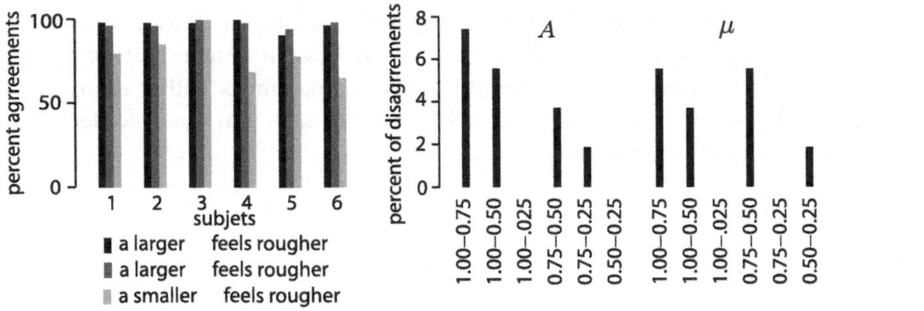

Fig. 8.2 Intra subject success rate of hypothesis and error rate for A and l as function of their difference. The *left panel* shows the percentage of times that each subject (1–6) chose the rougher surface in agreement with the hypothesis. The *right panel* shows an histogram of the percentage of trials that do not agree with the monotonicity hypothesis. For example, in 7.5% of the comparisons between textures with $A = 1$ and $A = 0.75$ the second was reported rougher. On the contrary, surfaces with $A = 1$ were always perceived rougher than $A = 0.25$, in perfect agreement with the hypothesis

due to the aforementioned ambiguity of the notion of roughness. Only one subject, 3, consistently chose the finer texture as rougher.

Figure 8.2 shows also the distribution of disagreements as a function of the values being compared. Interestingly, subjects mostly disagreed with the hypothesis when asked to judge pairs of textures with high values of A and μ. More importantly, large differences in parameters were seldom contrary to the hypothesis: 1.00–0.25, 0.75–0.25, and 0.5–0.25 had at most 1 error over 54 trials; notice that a similar difference in parameters 1.00–0.50 was misjudged more often. These two results indicate that, even if monotonic, the psychometric function relating roughness with A and μ is probably not linear, because similar differences in parameters did not result in similar error rates.

8.7 Discussion and Conclusion

Our aim was to validate the hypothesis that there exist more than one parameter which, when increased, also increase the perception of virtual roughness. We employed a texture synthesis algorithm that has a straightforward physical interpretation. When one drags a stylus against an object, if the surface is smooth then the dry friction force does not vary. If the surface is not smooth, hence "rough", then this force varies. A natural way to parametrize this variation is to consider that the more a surface deviates from smoothness, the deeper these variations are. It can also be observed that the deeper the grooves are the greater tendency a stylus has to get caught in the crevices, which *on average*, may be represented by a greater coefficient of friction.

To test this hypothesis we engineered a hardware platform with which we are confident most known haptic synthesis artifacts were eliminated. It has (a) a non-structural response, (b) plenty of resolution, and (c) generates a provably passive mechanical stimulus (like a real surface) under all the needed testing conditions. With this hardware, the results do support our initial hypothesis that increasing either depth of modulation or the coefficient of friction increases monotonically the perception of virtual roughness. It can then be concluded that the underlying psychometric functions are also monotonic. This result was obtained without instructing the participants of what roughness was.

Turning our attention to the effect of the spatial period. It is indeed hard to imagine a simple connection between spatial period and deviation from smoothness, unless innumerable assumptions are made regarding the exact nature of the interaction between a virtual stylus and a virtual surface. It is probable that our volunteers because of their varied backgrounds spontaneously utilized different sets of assumptions to answer the question they were asked with respect to changes in spatial period.

Finally, the hypothesis, presented by Smith et al., that the rate of change in the lateral force correlates with roughness is consistent with our analysis of A and μ. Further study is needed, however, to confirm this hypothesis in the virtual world; complete psychometric functions of roughness with respect A and μ needs to be measured, and their joint influence on roughness needs to be investigated.

Acknowledgements This work was funded by a Collaborative Research and Development Grant "High Fidelity Surgical Simulation" from NSERC, the Natural Sciences and Engineering Council of Canada and by Immersion Corp., and by a Discovery Grant also from NSERC.

References

1. Campion, G., Hayward, V.: Fundamental limits in the rendering of virtual haptic textures. In: Proceedings of the First Joint Eurohaptics Conference and Symposium on Haptic Interfaces for Virtual Environments and Teleoperator Systems, WHC'05, pp. 263–270 (2005)
2. Campion, G., Wang, Q., Hayward, V.: The Pantograph Mk-II: A haptic instrument. In: Proceedings of the IEEE/RSJ International Conference on Intelligent Robots and Systems, IROS'05, pp. 723–728 (2005)
3. Colgate, J.E., Schenkel, G.: Passivity of a class of sampled-data systems: Application to haptic interfaces. In: Proceedings of the American Control Conference, pp. 3236–3240 (1994)
4. Gosline, A.H., Campion, G., Hayward, V.: On the use of eddy current brakes as tunable, fast turn-on viscous dampers for haptic rendering. In: Proceedings of Eurohaptics, pp. 229–234 (2006)
5. Hayward, V.: Haptic synthesis. In: Proceedings of the 8th International IFAC Symposium on Robot Control, SYROCO 2006, pp. 19–24 (2006)
6. Hayward, V., Armstrong, B.: A new computational model of friction applied to haptic rendering. In: Corke, P., Trevelyan, J. (eds.) Experimental Robotics VI. Lecture Notes in Control and Information Sciences, vol. 250, pp. 403–412 (2000)
7. Klatzky, R.L., Lederman, S.J.: Perceiving textures through a probe. In: Touch in Virtual Environments, pp. 180–193. Prentice Hall PTR, Upper Saddle River (2002). Chap. 10
8. Kornbrot, D., Penn, P., Petrie, H., Furner, S., Hardwick, A.: Roughness perception in haptic virtual reality for sighted and blind people. Percept. Psychophys. **69**(4), 502–512 (2007)
9. Lederman, S.J., Klatzky, R.L., Tong, C., Hamilton, C.: The perceived roughness of resistive virtual textures: II. Effects of varying viscosity with a force-feedback device. ACM Trans. Appl. Percept. **3**(1), 15–30 (2006)

10. Minsky, M., Lederman, S.J.: Simulated haptic textures: Roughness. In: Proceedings of the ASME IMECE Symposium on Haptic Interfaces for Virtual Environments and Teleoperator Systems, vol. DSC-Vol. 58, pp. 421–426 (1996)
11. Smith, A.M., Chapman, C.E., Deslandes, M., Langlais, J.S., Thibodeau, M.P.: Role of friction and tangential force variation in the subjective scaling of tactile roughness. Exp. Brain Res. **144**(2), 211–223 (2002)
12. Wall, S.A., Harwin, W.S.: Interaction of visual and haptic information in simulated environments: Texture perception. In: Proceedings of the 1st Workshop on Haptic Human Computer Interaction, 31st August–1st September, pp. 39–44 (2000)

Chapter 9
Calibration of Virtual Haptic Texture Algorithms

Abstract Calibrating displays can be a time-consuming process. We describe a fast method for adjusting the subjective experience of roughness produced by different haptic texture synthesis algorithms. Efficiency results from the exponential convergence of the "modified binary search method" (MOBS) to a point of subjective equivalence between two virtual haptic textures. The method was applied to calibrate the modulation of the normal interaction force component against modulating a tangential friction force component. A table establishing the perceptual equivalence between parameters having different physical dimensions was found by testing 10 subjects. The method is able to overcome significant individual differences in the subjective judgement of roughness because roughness itself never needs to be directly estimated. A similar method could be applied to other perceptual dimensions provided that the controlling parameter be monotonically related to a subjective estimate.

9.1 Preface to Chap. 9

This chapter presents a fast method for calibrating the roughness sensation generated by two haptic texture algorithms. This calibration method is used to find the set of parameters for which two specific texture algorithms generate an equivalent sensation of roughness. Once the match is attained, it is possible to evaluate fairly the two calibrated algorithm; for example, the passivity margins at the point of subjective equivalence can be evaluated with the characteristic number.

The analysis tools developed in the previous chapters of the book, ensure that the calibration experiment is free of artifacts.

The paper contains two elements of novelty: first, the friction based texture algorithm introduced in Chap. 4 is shown to elicit a perception of roughness comparable to the one generated by a geometry based method, confirming the validity of the new formulation. The second point of novelty regards the perception of roughness: results show a significant effect of the subjects over the point of subjective equivalence.

Reprinted from Gianni Campion and Vincent Hayward, "Fast Calibration Of Haptic Texture Synthesis Algorithms." *IEEE Transaction on Haptics*, Volume 2, Number 2, 85–93, 2009.

G. Campion, *The Synthesis of Three Dimensional Haptic Textures: Geometry, Control, and Psychophysics*, Springer Series on Touch and Haptic Systems,
DOI 10.1007/978-0-85729-576-7_9, © IEEE 2009

9.1.1 Contribution of Authors

Gianni Campion worked on all the aspects of the work and on the manuscript; Prof. Hayward contributed to the design of the experiment, analyzed the results and edited the manuscript.

9.2 Introduction

The function of displays is to reproduce those aspects of the ambient physics that can satisfy our perceptual system in the accomplishment of tasks. For example, reading text on a computer screen requires a minimum level of contrast [21]; color appreciation requires fine characterization of the device used to mediate the conversion from a signal representation to its physical realization [26]; and so on. In applications where the signal to be displayed is processed digitally, the algorithms responsible for the synthesis participate in the result as much as the transducers do. In the graphics and audio domains, methods have been proposed to assess and compare algorithms, see for example [24, 27].

Haptics engineers face similar problems [22]. While it is known that force-feedback devices of different makes can produce different perceptual results [11], to our knowledge, there has been no report of suitable methods to perceptually compare different haptic synthesis algorithms or to link the psychophysics of haptic perception to the properties of the algorithms employed. Among the gamut of rendering approaches [23], the synthesis of textures stands out because the input to an algorithm may be specified in a simple and unambiguous manner, i.e., a surface geometry. The methods for converting a geometry into a force field that a user can experience are nevertheless far from being unique.

Depending on the assumptions made, there are many different ways to convert a surface geometry into a force field. The same surface geometry can give rise to different fields according to the assumed geometry of a probe tip, as a first example. Besides geometry, many further assumptions must be made regarding the materials in contact, their compliance, the resulting tribology, the statics and dynamics of the probe, the presence of foreign particles, of lubricants, and so on.

During the course of experimentation, we noticed that small changes in a synthesis algorithm could yield tangible differences. This observation raised the question of how algorithms could be compared with regard to the perceptual experience that they produce. The main purpose of this paper is to describe a fast psychometric method able to calibrate a pair of haptic synthesis algorithms with respect to each other from the view point of one particular perceptual dimension. Efficiency is essential since a complete perceptual characterization of texture is complex [2]. Our study confirms that a texture algorithm based on non-geometrical cues produced by modulating the tangential friction force can elicit a perception of roughness equivalent to that given by a geometry-based algorithm.

The proposed method was applied here to a pair of texture synthesis algorithms so that they elicit a perceptually equivalent sensation of roughness, despite their dependency on different sets of parameters. It is then possible to compare them fairly for their intrinsic properties such as implementation complexity or range of sensations they can synthesize. In particular, it becomes possible to quantify the relative passivity margin that algorithms provide for an equivalent level of perceptual roughness or conversely, the relative roughness experienced for the same passivity margin.

9.3 Related Work

Works related to the perceptual quantification of texture synthesis algorithms include those of Ho et al. [12]. They compared the influence of two- and three-dimensional force fields on perception and concluded that a square wave rendered as a three-dimensional force field is indistinguishable from the two-dimensional version if either amplitude or spatial period of the texture was smaller than 1.5 mm. Weisenberger et al. studied human detection performance of orientation for two- and three-dimensional textures experienced through different devices [29]. The results support the notion that the only significant difference was between sinusoidal textures and square waves in three dimensions. The authors also compared viscous-based textures with spring-based textures, but, due to the limited data available, no significant difference was found. The question of the perceptual properties of specific algorithms was briefly discussed in [7].

9.4 Approach

As alluded to earlier, the perceptual calibration of a device or of an algorithm can be a time-consuming process. In general, calibration requires to cyclically adjust parameters until two perceptual outcomes become subjectively equivalent. It is a search process where the objective can be simply defined but where the search space grows combinatorially fast with the number of parameters involved. A powerful simplifying property of any search problem is to depend on parameters that are monotonically related to the objective. The monotonicity property guarantees the success of any gradient-based search, that is, in spite of the fact that adjusting one parameter may undo the effect of another, progress toward the goal is guaranteed. By analogy, in color displays, it is by far more convenient to adjust chromaticity using "hue, saturation, and value" parameters than using "red, green, and blue" parameters.

Haptic Roughness Studies Arguably, roughness is the dominant property that one may want to match when substituting one algorithm for another—or one device for another—in a virtual reality haptic simulation. Roughness has been the topic of many studies, using both natural and electromechanically-generated stimuli. A current theory proposes a dual coding underlying texture perception [13]. According to this theory, the perception of microtextures (spatial period smaller than 0.2 mm) would be determined by the vibration of the skin moving against a surface and roughness estimation would follow a perceptual model based on the sensitivity of the Pacinian system. For coarser textures, roughness estimation would depend on the spatial characteristics of a texture. Studies using raised dot patterns and gratings showed a strong influence of inter-element spacing, but have produced conflicting results [18, 25]. The U-shaped spacing-to-roughness function found when appreciating texture with a bare finger shows an inversion point for inter-element spacing of 3.5 mm. The shape of this function seems to be preserved when a probe is used [16], but the inversion point shifts according to the tip geometry and exploration speed, denoting a lack of perceptual invariance [17, 19, 20].

Individual Differences Multidimensional scaling analysis employed by Hollins and colleagues revealed that texture perception suffers from significant inter-subject variations [14]. Specifically, the perceptual space of some subjects could be best described with the two dimensions of roughness and compliance, while other subjects responded to a third factor, the sticky-slippery dimension. Studies often examined subjective roughness estimation and discrimination using stimuli that had similar geometries. Here, we used stimuli that are fundamentally different, one being based on geometry and the other on frictional properties. It could therefore be expected that inter-subject differences in texture perception would be exacerbated. Roughness plays a key role in texture perception, but little is known regarding the individual variations in roughness perception, since experimental designs are typically not aimed at detecting inter-subject variations. Yet, virtual environments must work equally well for most people, in spite of these variations. Again the analogy with color screens is apt. Today, whether screen displays use LCD technology, phosphorescence, or plasma cells, commercial devices have achieved a fair degree of perceptual equivalence despite great differences in the raw stimulus and individual preferences.

Precautions To ensure that the experiment was carried out in the best conditions possible, we first analyzed the minimum requirements which a haptic simulation system must meet to be able to generate predictable textural effects [3]. Secondly, we developed a systematic method to characterize the properties of texture synthesis algorithms for their control and mathematical properties [4]. Third, we picked a pair of texture synthesis algorithms selected for their differences in their approach used to convert surface shape to a force field. Fourth, we ascertained that sensation perception scaled monotonically with the value of these parameters using the same apparatus and algorithms as in the present study [6].

9.4.1 System Considerations

For calibration to be meaningful, one must ensure that programmed changes in parameters are reflected in what is felt by the user. Earlier, we described six inequalities which reflect the ability of a system to synthesize and reproduce given textures [3].

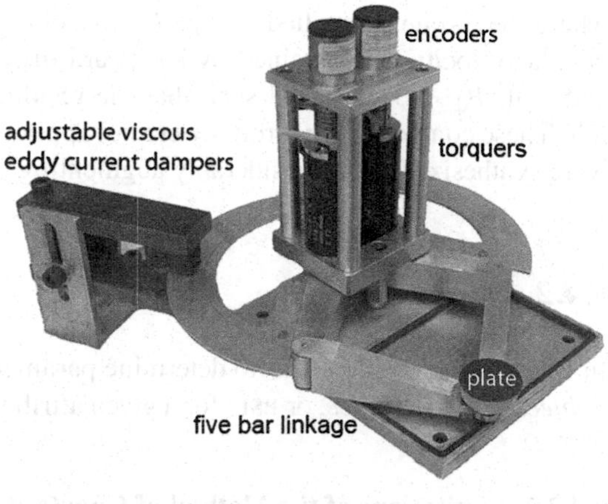

Fig. 9.1 Apparatus. The Pantograph device has a direct-driven five-bar mechanism that drives a small plate in the horizontal plane. In the present version, each joint was retrofitted with permanent magnet eddy-current viscous damper. The amount of viscosity is adjusted by controlling the amount of overlap between the polar pieces and the rotating annular blades

These conditions are not recalled here, but we ensured that they were met in our experiments. The fifth condition, however, namely that the product of the spatial frequency by the amplitude of a sinusoidal texture must be bounded by a constant, has immediate implications for roughness calibration. This trade-off means that for a given algorithm, in the course of an experiment not all combinations of parameters are valid. This inequality also raises the prospect that a pair of algorithms could never be calibrated for roughness, or that regions of achievable equivalent roughness are small.

To avoid the artifacts typical of texture synthesis, we used a highly reliable, custom-made hardware platform called Pantograph, which is fully described in Chaps. 3 and 4. This device meets all the required hardware-related requirements. It display forces of up to 2 N in a two-dimensional workspace of 100 × 60 mm; and within a flat bandwidth from DC to 400 Hz, see Fig. 9.1, making this device well described by a rigid body model without structural dynamics. The torque commands are processed by a low-pass reconstruction filter, so that the commands to the motors match the mechanical bandwidth of the system.

With the aim of further reducing possible artifacts and increasing the range of valid parameters, the device was retrofitted with viscous dampers based on the principle of eddy current brakes, see [9] for details. Two annular blades were concentrically mounted on the proximal arms so as to rotate through the air gap of C-shaped magnetic circuits. In the present configuration, a fixed magnetic field was generated by rare-earth magnets adjusted to give a damping coefficient, B_a, of approximately 6 mN m s. Interacting with the non powered Pantograph felt like exploring a viscous field.

To cancel the effects of this viscous field in free space, negative, generative damping was actively produced by the motors to cancel the inherent damping given by the dampers. This technique initially suggested by Colgate and Shenkel [8] was to our knowledge, for the first time successfully applied, thanks to the accurate viscous damping given by eddy current brakes. In each joint, the compensating torque was produced according to

$$\tau_k = +B_\mathrm{d}\hat{\omega}_k, \qquad (9.1)$$

where the torque τ_k applied at time k was computed from an estimate of the joint angular velocity, $\hat{\omega}_k$, obtained by backward difference over a 24-sample window, and with $B_d < B_a$ adjusted such that the residual viscosity field could barely be felt. These compensating torques were not applied inside the regions where textures were synthesized, thus considerably augmenting the system's passivity margin.

9.4.2 *Psychophysics*

In essence, the problem was to determine parameter settings that achieve a *Point of Subjective Equivalence*, or PSE, for a given attribute within a population of subjects.

9.4.2.1 Limitations of the Method of Constant Stimulus

It is well accepted that an accurate method to determine a PSE is to measure complete psychometric functions by the method of constant stimulus. In general, a psychometric function relates the value of a single parameter to the probability of a perceptual event to occur, such as the detection of a weak signal or discrimination between different signals. Once such function is known, the value of the parameter corresponding to the 50% probability determines the point of subjective equivalence. To measure PSE's, we could present subjects with two textures, one synthesized by a reference algorithm and one synthesized by the algorithm to be calibrated. Subjects would report which of the two virtual surfaces felt rougher and parameters would be adjusted until the reference surface felt rougher 50% of the times. With the method of constant stimuli, the psychometric curve should be sampled at least at five locations. To obtain statistically significant results, the number of trials needs to be large, in the order of a few hundreds.

The method of constant stimuli is often preferred to determine the behavior of a population in a scientific study where confounding factors like boredom and sensory adaptation can be controlled. Then, the assumption that the random process being sampled is stationary can be valid. In the simplest case, a perceptual outcome depends on at least two parameters, in the present case, we must vary at least amplitude and spatial frequency. A bare bones experiment would require to test for at least three values of each parameter, thereby forcing users to sit through thousands of trials. This approach is not viable when the objective is to calibrate a machine, possibly specifically for each individual user.

9.4.2.2 Advantages of Staircase Methods for Calibration

Another approach is to employ staircase methods because they can be adapted to finding specific values of the psychometric function, not the whole function. Instead of uniformly sampling the psychometric function, most of the trials are concentrated in the neighborhood of a target value. For example, the common 1-up 1-down method, converges to the value corresponding to the 50% probability. Even faster results can be obtained with adaptive methods, to further reduce the sampling requirement. Staircase methods (both with fixed and adaptive step size) do not allow for a reliable reconstruction of the complete psychometric function because its graph is coarsely sampled [15]. This property, instead of being a drawback, is an advantage for our present purpose.

9.4.2.3 Modified Binary Search Method

Among the family of adaptive staircase methods, the "MOdified Binary Search" method, or MOBS [28], is designed to converge exponentially to the 50% point of the psychometric function using an adaptive step size. It relies on binary search, while accounting for threshold migration during an experiment by using multiple stacks. It was originally designed for visual contrast detection and can be adapted to yes/no experimental designs. The main reasons for selecting this method are its speed and simplicity. MOBS has also a characteristic which is highly relevant to haptic texture calibration. The limitations of any haptic device, such as resolution and damping, create tradeoffs between the parameters of the textures that can be reliably synthesized without artifacts. This problem impacts any thresholding process since a sought threshold can fall outside the range of admissible parameters. MOBS can cope with this occurrence because it keeps tracks of multiple candidate intervals and because its fast convergence allows for repetitions of the same thresholds. Finally, MOBS' performance is at par with more complex and rigorous thresholding approaches [1]. Taken together, these properties have advantages that greatly outweigh those of other methods.

9.4.2.4 Brief Description

The first step of MOBS is to define an interval that should contain the value of the parameter to be thresholded. The experiment proceeds by testing the midpoint of this interval, which is then reduced by dichotomy based on the answer of the subject. To account for threshold migration, the algorithm tests the boundary of the interval when the subject does not reverse her or his answer twice in a row. If the threshold falls outside the current interval, the method rolls back to a previous boundary. At any time, a history of 3 upper and 3 lower bounds is kept in two stacks; the update rules for the stacks can be found in [28]. The termination criterion is based on a fixed number of reversals as well as on the size of the interval. If a prescribed number of reversals is reached but the interval is too large, the trial is extended for two more reversals; when the trial ends, the estimated threshold is the midpoint of the last interval. For a complete description please refer to [1].

9.5 Materials and Methods

We selected two texture synthesis algorithms that were as different in concept as possible. We then asked subjects to calibrate one algorithm against the other using

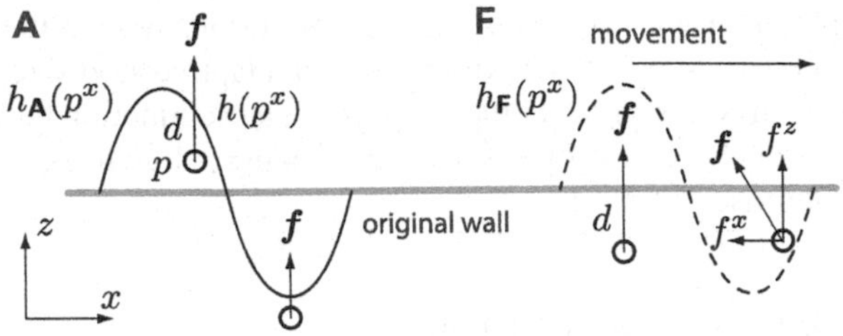

Fig. 9.2 Algorithm **A** produces a force field aligned with the z direction and with a magnitude that is proportional to the penetration measured along z from the undulating boundary. Algorithm **F** produces a field resulting from a normal component resulting from penetrating a smooth wall and a lateral component resulting from a friction force modulated as a function of the lateral displacement

the approach just described to find parameters that produced an equivalent sensation of roughness. In doing so, we never asked subjects to directly estimate roughness with either of the two algorithms. At the conclusion of the experiment it was possible to find such parameters, but other aspects of the texture synthesis differed between the two algorithms.

9.5.1 Algorithms

Textured walls were created in two dimensions, x, z. The point $p = [p^x, p^z]^\top$ represents the position of the probe in the virtual space. A geometrical profile $h(p^x)$ was applied to a straight virtual wall of stiffness κ_0. Please refer to Fig. 9.2 for an illustrative description of the algorithms.

Algorithm **A** produces a force field according to:

$$f_{\mathsf{A}}(p) = \begin{cases} -\kappa_0[0, (p^z - h_{\mathsf{A}}(p^x))]^\top & \text{if } (p^z - h_{\mathsf{A}}(p^x)) < 0, \\ [0, 0]^\top & \text{otherwise,} \end{cases}$$
$$h_{\mathsf{A}}(p^x) = A \sin(2\pi p^x / L), \tag{9.2}$$

where penetration is computed from the boundary of the texture along the z direction, A is the amplitude and L is the spatial period. Algorithm **F** is based on the modulation of dry friction to generate a texture. A time-free friction model, described in [10], is used to generate lateral force component which is then modulated according to the height function $h_{\mathsf{F}}(p^x)$. The force field is

$$f_{\mathsf{F}}(p) = \begin{cases} -\kappa_0[\mu[1 - h_{\mathsf{F}}(p^x)]\frac{d^x}{d^x_{\max}} p^z, p^z]^\top & \text{if } p^z < 0, \\ [0, 0]^\top & \text{otherwise,} \end{cases}$$
$$h_{\mathsf{F}}(p^x) = \sin(2\pi p^x / L), \tag{9.3}$$

where d^x is the pre-sliding tangential displacement, $d^x_{\max}$ its the maximum value (1 mm), μ is Amontons' coefficient of friction. An additional parameter, $d^z_{\max}$

(0.5 mm), is introduced to limit the value of p^z in the computation of the friction force.

The most notable difference lies in the energetic properties of respective force fields given by the algorithms. Neither field is conservative; algorithm **A** produces a generative force field, while **F** gives a field that is generally dissipative.

9.5.2 Characteristic Number

The algorithms under investigation were thoroughly analyzed in [4]. From this reference, two notions are directly relevant to the present study. The first notion is that if an algorithm generates a non-conservative field, then the resulting synthesis can be tainted by artifacts even if a haptic force-feedback system is locally passive. The second is the concept of a characteristic number for a synthesis algorithm. When a virtual wall of stiffness κ_w is textured using a given algorithm, the system passivity margin changes according to the norm of the Jacobian matrix of the force field generated by the virtual environment, $\kappa_t = \|J_{ve}\|_2$. The characteristic number of an algorithm is the ratio $q = \kappa_t/\kappa_w$, which represents the change in passivity margin due to "painting" a texture on a smooth surface. In most cases, the characteristic number is independent from the stiffness of the underlying virtual wall.

9.5.3 Experiment Design

The parameters of algorithm **A** were kept fixed when searching for a perceptually equivalent value of parameter μ in algorithm **F**, and only textures having the same spatial frequencies were compared. This procedure avoided the issues related to the possible lack of monotonicity of the relationship spatial-period/roughness. The first step was to define a set of reference textures to be rendered with algorithm **A**. The characteristic number of this algorithm is of the form $q_\mathsf{A} = \sqrt{1+[2\pi A/L]^2}$ (see Appendix for detail). As a consequence, not all the combinations of A and L are admissible hence we limit our stimuli to $A/L \le 1$. Selecting five different values for the spatial period, $L = 0.12$, 0.25, 0.50, 1.00, and 2.00 mm gives fives values for parameter A as listed in Table 9.1, allowing fifteen valid combinations. This constraint can intuitively be understood by considering the shape of an undulating profile of fixed height. The smaller is the spatial period, the greater becomes the slope, and the greater becomes the 'control gain'. The larger values of q_A, particularly when $q_\mathsf{A} > 2$, indicate the cases where non passive behaviors are possible. The characteristic number, however, like other passivity-based measures, is conservative and the parameter value combinations on the diagonal happened to be acceptable in practice. These system constraints caused the experimental design to be unavoidably unbalanced.

The second step was to define the initial intervals for μ to be used with the MOBS thresholding method. The intervals were chosen so that the friction algorithm

Table 9.1 Characteristic number $q_{\mathbf{A}}$ for the reference textures

A (mm)	L (mm)				
	0.12	0.25	0.50	1.00	2.00
0.12	6.4	3.3	1.9	1.3	1.0
0.25		6.4	3.3	1.9	1.3
0.50			6.4	3.3	1.9
1.00				6.4	3.3
2.00					6.4

Table 9.2 Starting intervals for MOBS estimation of μ

	L (mm)				
	0.12	0.25	0.50	1.00	2.00
top	0.2	0.4	0.6	0.8	1.0
bottom	0	0	0	0	0

would be marginally non passive when rendering the highest possible value of μ in the interval, see Table 9.2.

For accurate calibration, seven reversals were required and the terminal interval was to be 1% of the size of the inital interval, otherwise the trial was extended for two additional reversals. As already mentioned, a newly introduced termination criterion was needed because, due to passivity constraints, the initial interval could not be made arbitrarily large and therefore was not guaranteed to contain the subjectively equivalent threshold value for μ. To deal with this problem, the trial was ended if the three upper boundaries and the three lower boundaries in the interval stacks were all equal; the threshold was then set to the value of the boundary.

Because intervals cannot be made arbitrarily large, some resulting threshold values were clipped and did not reflect a true threshold. This occurrence was accounted for in the analysis of the results, where order-based tests were preferred. In particular, the median was used to indicate the central tendency of the samples.

9.5.4 Subjects and Experimental Procedure

A total of 10 paid subjects participated in the experiment (3 male and 7 female). Although the Pantograph device operates quietly, a faint noise sometimes emanates from the actuators, which may taint the results. Subjects were asked to wear sound isolation headphones (DirectSound™ EX-29) through which white noise was played at a self-adjusted volume. The apparatus was concealed behind a curtain.

Subjects interacted with the Pantograph by putting the index finger of their dominant hand on the Pantograph's plate. An experiment consisted of a sequence of MOBS thresholding trials, three for each of the 15 reference textures. Subjects were presented with two textures synthesized on two parallel facing virtual walls, 30 mm

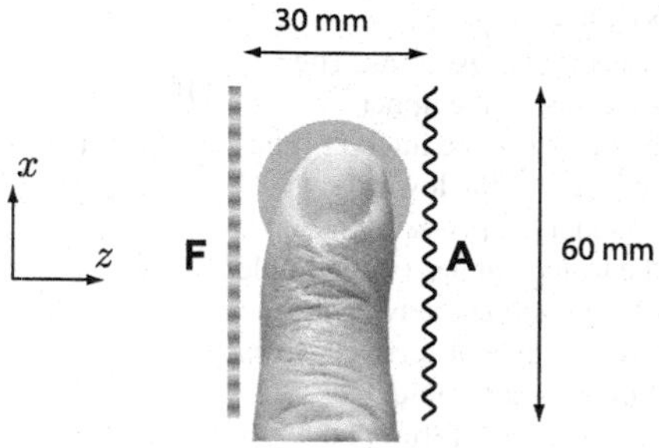

Fig. 9.3 Experiment layout. The two textured virtual walls were 30 mm apart, randomly disposed on the left or on the right throughout the trials

apart. See Fig. 9.3. The reference texture given by algorithm **A** was compared with a test texture given by algorithm **F**.

Subjects were asked to decide which of the two textures was rougher, and their answers were entered via keyboard strokes. After each answer, the state of the MOBS procedure was updated, a new μ was computed and presented to the subject. This sequence was repeated until convergence, which was typically achieved in less than two minutes.

Subjects were given a single training trial, to familiarize them with the haptic device and with the movements required for exploration. No feedback was given. Subjects were instructed to proceed as fast as possible but no restriction was imposed on their exploratory procedure. They could explore the surfaces in any order and as many times they felt it necessary to reach a judgment.

Reference textures were presented to each subject in a randomized order. The order was the same for all subjects to facilitate analysis. The objective was to avoid bias in the subject's responses. Due to the adaptive nature of the experiment (and hence a variable number of presentations) and because of the lack of exploratory constraints (different exploration time for each of the trials), we did not expect a significant contribution of learning effects in the resulting thresholds. The reference texture was presented either left or right with a 50% probability. Each subject estimated 3 thresholds for μ with each of the 15 reference textures. At the end of the experiment, 450 thresholds were recorded. To avoid biases, subjects *were not instructed* as to what was meant by roughness. To avoid confusion, "roughness" was simply defined as being "the opposite of smoothness".

9.6 Results

9.6.1 Raw Data

Subjects tended to produce four patterns of convergence, described as follows.

- Perfect convergence was characterized by an exponentially contracting interval resembling an overdamped second-order response, until the required number of reversals was achieved, see Fig. 9.4(a).

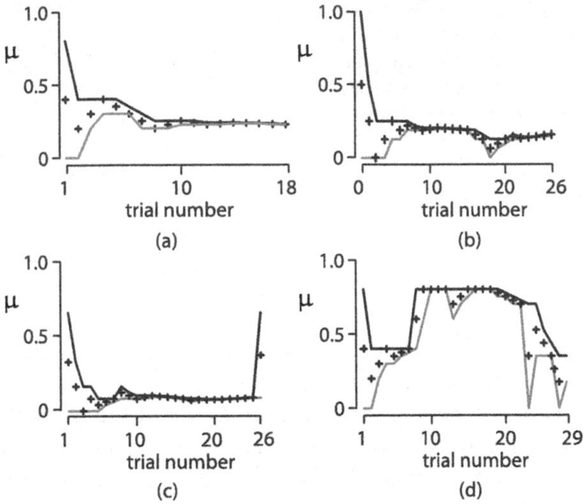

Fig. 9.4 Types of convergence patterns. The *solid line* is the upper boundary of the interval. The *grey line* is the lower boundary. *Crosses* represent the tested values. (**a**) Example of exponential convergence. (**b**) Example of recovery from a drifting threshold. (**c**) Example of trumpet artifact which occurred in 1% of cases. (**d**) Rare example of lack of convergence

- 'One-bump' convergence occurred when the threshold drifted sufficiently to force a reset to an initial boundary value. MOBS can still converge after a reset if there is a sufficient number of reversals left in the trial, Fig. 9.4(b). Some trials exhibit multiple bumps before convergence.
- A 'trumpet artifact' occurred when the threshold drifted toward the end of a trial. In response, the interval had to increase dramatically near the end of a trial, see Fig. 9.4(c). MOBS cannot recover from such occurrence. These artifacts were observed in 1% of the trials and were corrected for by discarding the last 2–3 answers.
- Lack of convergence. In approximately 5% of all trials, subjects were not able to consistently compare textures, see Fig. 9.4(d). These cases were not discarded and were considered as noise in the calibration procedure.

Overall, the method was found to converge to a clear threshold, directly or with "bumps", in 94% of all trials. After correcting for the "trumpet" artifacts, the convergence rate was about 95%.

Repeating the thresholding process three times made failure to converge very unlikely. In only one case out of 150 trials two of the three thresholds were not perfectly convergent, but since the two were consistent with each other no action was taken. The few cases in which MOBS did not converge had no significant influence on the results. The most notable artifact that can be attributed to MOBS results from an exaggerated modification of the intervals in response to drifting thresholds.

9.6.2 Analysis of the Overall Results

Please refer to Fig. 9.5 for a plot of the distribution of the 450 estimated threshold values sorted by amplitude of the reference texture.

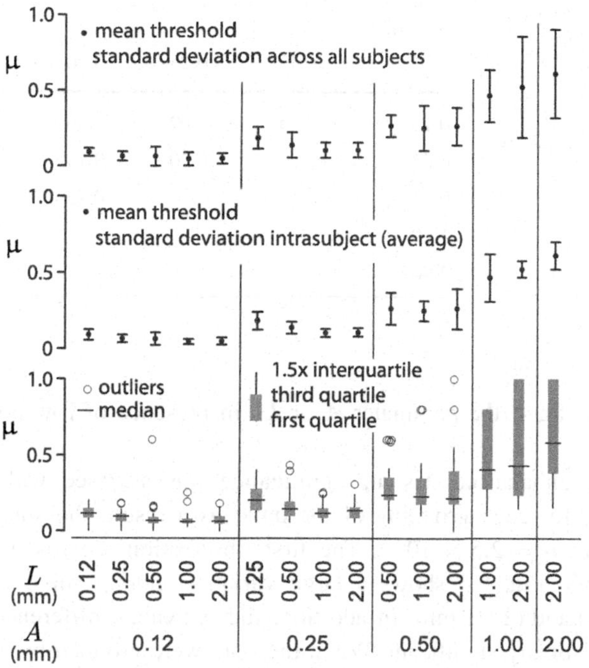

Fig. 9.5 Results of calibration procedure where 450 thresholds were estimated. The *bottom panels* shows the statistical distribution of the thresholds. The other panels present the standard deviation intra and among subjects

The estimated threshold of μ for the point of subjective equivalence was strongly correlated with the amplitude A in algorithm **A**. The Spearman correlation test gave $\rho = 0.7889$ ($n = 450$, $p < 10^{-6}$) and the Pearson correlation test gave $r = 0.6980$ ($n = 450$, $p < 10^{-6}$), showing a strong and significant monotonic linear correlation between these quantities. Also, the median values of μ (over all subjects and thresholds) correlated linearly with the amplitude A, $r = 0.9661$ ($n = 15$, $p < 10^{-6}$).

In all, the Pearson and Spearman correlation tests indicated a strong and significant linear correlation between the parameter μ of algorithm **F** and the parameter A of algorithm **A**. Moreover, the transformation A [**A**] $\leftrightarrow$ roughness $\leftrightarrow$ μ [**F**] was monotonic and largely linear. This strong parameter correlation is an important result and was obtained without explicitly relating the parameters to roughness.

Due to the limited number of data points per reference texture and because of the limited size of the initial intervals, the significance of the data was assessed through Friedman tests. The triangular size of the data forced us to use eight different tests, performed on the estimated thresholds grouped first by spatial frequency and then by amplitude. Subsequently, repeated Friedman tests were performed to assess pairwise significant differences. Subjects were treated as random row factors with three repetitions, by using a two-way Friedman test, for the significance

Table 9.3 Medians of the estimated thresholds

A (mm)	L (mm)				
	0.12	0.25	0.50	1.00	2.00
0.12	0.11	0.09	0.06	0.06	0.06
0.25		0.20	0.14	0.12	0.13
0.50			0.23	0.23	0.22
1.00				0.40	0.43
2.00					0.58

of the column effect (the parameter A or L) in presence of row effect (the subject).

Due to the unbalanced design, significance was assessed with 20 pairwise Friedman tests for each grouping of the data; as a result, the significance level was lowered to $\alpha = 2.5 \times 10^{-3}$. The first observation regards the distribution of the raw data; Fig. 9.5 shows a large variation among subjects for textures $(A, L) = (1, 2)$ and $(2, 2)$ mm. In addition, the individual differences of the median were statistically significant. When the data were grouped by amplitude, the spatial frequency showed significant effect for $A = 0.12$ mm, $(p < 2 \times 10^{-5})$ and $A = 0.25$ mm, $(p < 2 \times 10^{-4})$. Repeated pairwise Friedman tests suggest that textures with $L \leq 0.25$ mm lead to significantly different estimations of μ than textures with $L > 0.25$ mm. Interestingly, this partition is similar to the distinction made between micro and macro textures [13].

When the data are grouped by spatial period, we should expect significant differences. In fact, the distribution of μ showed significant differences due to amplitude ($p < 10^{-6}$); the median values of μ for each reference texture are reported in Table 9.3. Further investigations with repeated pairwise Friedman tests showed significant pairwise differences between textures with the same spatial period ($p < 2.5 \times 10^{-3}$). The only exception is the pair $A = 1, 2$ mm, $L = 2$ mm for which significance was not found ($p > 6 \times 10^{-3}$). These results confirm that the calibration procedures generally assigns a significantly different μ estimate for each different A.

Using the results of Table 9.3 we are now in a position to compute the characteristic numbers of algorithm **F** obtained for an equivalent sensation of roughness compared to that given by algorithm **A**. The results are collected in Table 9.4. By comparing with the values in Table 9.1, algorithm **F** can be said to be on average 30% more passive than algorithm **A** for an equivalent sensation of roughness. The subjects in this sample were found to be a significant random factor for the estimates of μ, thus the passivity margins should be assessed on a per subject basis, which further motivates the need for a rapid calibration procedure. The mean values of the passivity margins are nonetheless reported in Table 9.4.

Table 9.4 Characteristic number of algorithm **F** for equivalent perceptual roughness

A (mm)	L (mm)				
	0.12	0.25	0.50	1.00	2.00
0.12	6.10	2.71	1.12	1.01	1.01
0.25		5.87	2.42	1.25	1.12
0.50			3.92	2.38	1.63
1.00				4.23	3.20
2.00					4.32

9.7 Discussion and Conclusion

In this paper we have described a method that can be used for fast perceptual calibration between haptic synthesis algorithms. Given careful characterization of the mathematical and control properties of algorithms as well as of the hardware platform used to transduce the signal, it was possible to show that a pair of algorithms operating on different principles could be calibrated to produce an equivalent roughness for a large range of parameter settings. Efficiency resulted from the use of a fast matching method operating from the principle of dichotomy search, which was adapted to the needs of the determination of the point of subjective equivalence.

Our main results are Tables 9.3 and 9.4. It can be seen from Eqs. (9.2) and (9.3) that each algorithm depends on two parameters where the spatial period, L, is common to both. The amplitude, A, is particular to the algorithm **A**, while **F** depends on the coefficient of friction, μ. It is therefore apparent that the two algorithms operate on different principles—A and μ have different dimensions—yet, they can be perceptually calibrated against each other.

Once this result was obtained, we could relate the two algorithms from the view point of the passivity margin that they provide. Specifically, an algorithm based on the principle of lateral friction force modulation, **F**, was found to be more passive than an algorithm based on perturbing the virtual interacting point in a direction orthogonal to the surface of a virtual surface, **A**. Conversely, algorithm **F** can provide a greater sensation of roughness than **A** for the same passivity margin. It is anticipated that the data reported in Table 9.3 would also be useful when the algorithms are employed with different hardware platforms since we have used a device having the most exacting characteristics.

The friction coefficient, μ, that drives the generation of non-geometric cues could be made to play a role equivalent to that of texture geometric profile amplitude, A. This result was obtained without an explicit estimation of roughness. In respect to the lack of a precise definition for the notion of surface roughness, some subjects interpreted reference textures with large spatial period $((A, L) = ((1, 2), (2, 2))$ mm) as being very rough, while others identified them as very smooth wavy surfaces. Assigning roughness to a low spatial frequency texture is difficult, which is reflected by significant variance across subjects.

Acknowledgements The authors would like to thank Andrew H.C. Gosline for the engineering of the eddy current brakes, Maarten W.A. Wijntjes and Ilja Frissen for advice with psychometric techniques. This work was funded by a Collaborative Research and Development Grant "High Fidelity Surgical Simulation" from the Natural Sciences and Engineering Council of Canada (NSERC), and by Immersion Corp. Additional funding is from a Discovery Grant from NSERC for the second author.

Appendix: Characteristic Numbers

9.8.1 Algorithm A

The Jacobian matrix of the force field is

$$\mathbf{J}_{f\mathsf{A}}(p) = -\kappa_0 \begin{bmatrix} 0 & 0 \\ -h'_{\mathsf{A}}(p^x) & 1 \end{bmatrix}. \tag{9.4}$$

Its norm is

$$\|\mathbf{J}_{f\mathsf{A}}\|_2 = \kappa_{\mathsf{A}} = \kappa_0 \sqrt{1 + [h'_{\mathsf{A}}(p^x)]^2} \tag{9.5}$$

which gives

$$q_{\mathsf{A}} = \max \kappa_{\mathsf{A}}/\kappa_0 = \sqrt{1 + [2\pi A/L]^2} \tag{9.6}$$

when $h_{\mathsf{A}}(p^x) = A\sin(2\pi p^x/L)$.

9.8.2 Algorithm F

The Jacobian matrix of the force field is

$$\mathbf{J}_{f\mathsf{F}}(p) = -\kappa_0 \begin{bmatrix} \mu \frac{p^z}{d^x_{\max}} (\frac{\mathrm{d}d^x}{\mathrm{d}p^x} - h_{\mathsf{F}}(p^x)\frac{\mathrm{d}d^x}{\mathrm{d}p^x} - h'(p^x)) & \mu[1 - h_{\mathsf{F}}(p^x)]\frac{d^x}{d^x_{\max}} \\ 0 & 1 \end{bmatrix}. \tag{9.7}$$

In the worst case and according to [10], (9.7) becomes:

$$\mathbf{J}_{f\mathsf{F}}(p) = -\kappa_0 \begin{bmatrix} 2\mu p^z(1/d^x_{\max} + \pi/L) & 2\mu \\ 0 & 1 \end{bmatrix}. \tag{9.8}$$

For $h_{\mathsf{F}}(p^x) = \sin(2\pi p^x/L)$, $q_{\mathsf{F}} = \max \|\mathbf{J}_{f\mathsf{A}}\|_2/\kappa_0$ can be quickly numerically computed since the value of p^z must be clamped to a maximum $d^z_{\max}$.

References

1. Anderson, A.J., Johnson, C.A.: Comparison of the ASA, MOBS, and ZEST threshold methods. Vis. Res. **46**(15), 2403–2411 (2006)

2. Bergmann-Tiest, W.M., Kappers, A.M.L.: Analysis of haptic perception of materials by multidimensional scaling and physical measurements of roughness and compressibility. Acta Psychol. **121**, 1–20 (2006)
3. Campion, G., Hayward, V.: Fundamental limits in the rendering of virtual haptic textures. In: Proceedings of the First Joint Eurohaptics Conference and Symposium on Haptic Interfaces for Virtual Environments and Teleoperator Systems, WHC'05, pp. 263–270 (2005)
4. Campion, G., Hayward, V.: On the synthesis of haptic textures. IEEE Trans. Robot. **24**(3), 527–536 (2008)
5. Campion, G., Wang, Q., Hayward, V.: The Pantograph Mk-II: A haptic instrument. In: Proceedings of the IEEE/RSJ International Conference on Intelligent Robots and Systems, IROS'05, pp. 723–728 (2005)
6. Campion, G., Gosline, A.H., Hayward, V.: Does judgement of haptic virtual texture roughness scale monotonically with lateral force modulation? In: Proceedings of Eurohaptics. LNCS, vol. 5024, pp. 718–723. Springer, Berlin (2008)
7. Choi, S., Tan, H.Z.: Perceived instability of virtual haptic texture. III. Effect of update rate. Presence **16**(3), 263–278 (2007)
8. Colgate, J.E., Schenkel, G.: Passivity of a class of sampled-data systems: Application to haptic interfaces. In: Proceedings of the American Control Conference, pp. 3236–3240 (1994)
9. Gosline, A.H.C., Hayward, V.: Eddy current brakes for haptic interfaces: Design, identification, and control. IEEE/ASME Trans. Mechatron. **13**(6), 669–677 (2008)
10. Hayward, V., Armstrong, B.: A new computational model of friction applied to haptic rendering. In: Corke, P., Trevelyan, J. (eds.) Experimental Robotics VI. Lecture Notes in Control and Information Sciences, vol. 250, pp. 403–412 (2000)
11. Hayward, V., Astley, O.R.: Performance measures for haptic interfaces. In: Giralt, G., Hirzinger, G. (eds.) Robotics Research: The 7th International Symposium, pp. 195–207. Springer, Heidelberg (1996)
12. Ho, P.P., Adelstein, B.D., Kazerooni, H.: Judging 2D versus 3D square-wave virtual gratings. In: Proceedings of the 12th International Symposium on Haptic Interfaces for Virtual Environment and Teleoperator Systems, pp. 176–183 (2004)
13. Hollins, M., Bensmaia, S.J.: The coding of roughness. Can. J. Exp. Psychol. **61**(3), 184–195 (2007)
14. Hollins, M., Bensmaïa, S.J., Karlof, K., Young, F.: Individual differences in perceptual space for tactile textures: Evidence from multidimensional scaling. Percept. Psychophys. **62**(8), 1534–1544 (2000)
15. Kaernbach, C.: Slope bias of psychometric functions derived from adaptive data. Percept. Psychophys. **69**(8), 1389–1398 (2001)
16. Klatzky, R.L., Lederman, S.J.: Tactile roughness perception with a rigid link interposed between skin and surface. Percept. Psychophys. **61**(4), 591–607 (1999)
17. Klatzky, R.L., Lederman, S.J., Hamilton, C., Grindley, M., Swendsen, R.H.: Feeling textures through a probe: Effects of probe and surface geometry and exploratory factors. Percept. Psychophys. **65**, 613–631 (2003)
18. Lawrence, M.A., Kitada, R., Klatzky, R.L., Lederman, S.J.: Haptic roughness perception of linear gratings via bare finger or rigid probe. Perception **36**(4), 547–557 (2007)
19. Lederman, S.J., Klatzky, R.L., Hamilton, C.L., Ramsay, G.I.: Perceiving roughness via a rigid probe: Psychophysical effects of exploration speed and mode of touch. Haptics-E: Electron. J. Haptics Res. **1** (1999), online
20. Lederman, S., Klatzky, R., Hamilton, C., Grindley, M.: Perceiving surface roughness through a probe: Effects of applied force and probe diameter. In: Proceedings of the ASME DSCD-IMECE (2000)
21. Legge, G.D., Parish, D.H., Luebker, A., Wurm, L.H.: Psychophysics of reading. XI. Comparing color contrast and luminance contrast. J. Opt. Soc. Am. A **7**(10), 2002–2010 (1990)
22. Leškovský, P., Cooke, T., Ernst, M.O., Harders, M.: Using multidimensional scaling to quantify the fidelity of haptic rendering of deformable objects. In: Proceedings of Eurohaptics, pp. 289–295 (2006)

23. Lin, M., Otaduy, M. (eds.): Haptic Rendering: Foundations, Algorithms and Applications. A. K. Peters, Ltd, Wellesley (2008)
24. Seuntiens, P., Meesters, L., Ijsselsteijn, W.: Perceived quality of compressed stereoscopic images: Effects of symmetric and asymmetric jpeg coding and camera separation. ACM Trans. Appl. Percept. **3**(2), 95–109 (2006)
25. Smith, A.M., Chapman, C.E., Deslandes, M., Langlais, J.S., Thibodeau, M.P.: Role of friction and tangential force variation in the subjective scaling of tactile roughness. Exp. Brain Res. **144**(2), 211–223 (2002)
26. Stokes, M., Fairchild, M.D., Berns, R.S.: Precision requirements for digital color reproduction. ACM Trans. Graph. **11**, 406–422 (1992)
27. Tolonen, T., Järveläinen, H.: Perceptual study of decay parameters in plucked string synthesis. In: Proceedings of the 109th Convention of Audio Engineering Society. Preprint no. 5205 (2000)
28. Tyrrell, R.A., Owens, D.A.: A rapid technique to assess the resting states of the eyes and other threshold phenomena: The modified binary search MOBS. Behav. Res. Meth. Instrum. Comput. **20**(2), 137–41 (1988)
29. Weisenberger, J.M., Kreier, M.J., Rinker, M.A.: Judging the orientation of sinusoidal and square-wave virtual gratings presented via 2-DOF and 3-DOF haptic interfaces. Haptics-e **1**(4) (2000), online

Chapter 10
Conclusions

Abstract The book is concluded by a chapter summarizing the major findings and showing the possible extensions to the research presented in the previous chapters.

10.1 Summary

This book is the first attempt to formalize the specific artifacts corrupting the rendering of virtual haptic textures. At a first glance, this document can be read as a practical guide for precise haptic textures; this is partially the intent of the author: to offer a set of simple conditions to guide haptic researchers towards artifact-free textures. The conditions identified in this work are also extremely valuable when designing psychophysical experiments (because not all the textures can be rendered on a haptic device) and when analyzing the significance of the data collected.

10.2 Results

The guidelines of this book, however, are clearly motivated and, for the most part, experimentally validated; moreover, the passivity conditions and the characteristic number, required a novel interpretation of the same idea of passivity when applied to multidimensional virtual environments.

10.2.1 Passivity

The characteristic number is a measure of the impedance of virtual haptic texture algorithms, and, coupled with the novel interpretation of passivity, prevents control related artifacts. Alternative approaches are available in the literature, among the others, virtual coupling could stabilize the interaction, both for conservative and non-conservative force fields [1]. Another viable solution is the passivity observer with energy following, which, in real time, ensures the energy balance of the haptic interaction [3].

G. Campion, *The Synthesis of Three Dimensional Haptic Textures: Geometry, Control, and Psychophysics*, Springer Series on Touch and Haptic Systems,
DOI 10.1007/978-0-85729-576-7_10, © Springer-Verlag London Limited 2011

These on-line approaches, however, do not offer any insight on the nature of the unstable texture, hence their application could remove perceptually significant aspects of the textured force field. On the other hand, the characteristic number matches the maximum impedance renderable by the haptic device with the impedance of the virtual textures, which leads to more deterministic force fields, better suited for psychophysical studies.

10.2.2 Devices and Algorithms

The combination of the Pantograph and the new friction based algorithm is a well characterized experimental setup for the study of the perception of textures.

The oversample and filter approach is pivotal for the quality of the haptic textures rendered with the Pantograph, but alternative solutions exist. The acceleration matching technique, successfully applied to the PHANTOM™, imposes precise open loop acceleration profiles to the handle of the device, once the dynamic model of the combination device/user is identified. At this stage of development, acceleration matching seems to be better suited for rendering time-varying stochastic textures, because its applicability to closed-loop multidimensional virtual environments has not yet been investigated [4].

The most evident limitation of the Pantograph is the 2D workspace and the 2D forces it generates. Among the multidimensional devices, however, no device offers the same level of resolution, bandwidth, low inertia, and low friction. The only possible exception is the Ministick, whose frequency response, though, has not yet been measured [2].

The proposed friction based algorithm is physically inspired and plausible, and easily extendable to 3D curved surfaces. In addition, it has a number of parameters to fine tune the texture sensation and the passivity of the interaction, and its dissipative nature is instrumental for the rendering of a stable texture. Friction maps recorded from the surface of real objects, e.g., [5], can be rendered with this algorithm without modifications; moreover, an analysis of the gradient of those maps would ensure passive rendering. Finally, the extension to 3D objects of friction based approaches has a clear advantage over the geometry based methods, because the collision detection and the computation of the minimal distance are performed with the low curvature surface and not with the texture boundaries. This simplification is important, for example, when the applied texture is a profile measured from a real surface, because the collision detection algorithm is not affected by the potential complexity of the profile.

The innovations of this work are clearly targeted to the most demanding audience in the haptic community, those researchers who want to generate repeatable stimuli for the study of psychophysics of touch. The same techniques, however would benefit also the less demanding haptic applications: the parameters of the novel texture algorithm can be tweaked to account for the specification of most haptic devices; moreover, the characteristic number does apply to any texture algorithm regardless of the quality of the haptic device.

10.3 Future Work

The most enticing application of the work presented in this book is the psychophysics study of indirect touch. The innovations presented here finally permit to generate textural stimuli with guaranteed quality, ideal for the study of perceptual properties of virtual textures, among which roughness is of great importance. For example, it is possible to apply the conditions identified here to geometry profiles sampled from real surfaces, for the study of the perception of real versus virtual textures.

A second possible avenue of extension of this work, regards the application of haptic textures to virtual reality. The novel formulation of the friction based algorithm and, its three dimensional extension, would be the ideal candidate for adding textures to generic curved surfaces, without affecting the underlying computational engine. For example, in a surgical simulator, the texture of a virtual bone could be changed to express different states of decay of the tissues, without the need of a complex underlying geometry. On the other hand, the surfaces are in general represented with triangular meshes whose discontinuities are known to generate perceptual artifacts. These discontinuities are usually handled by "blending" the normals of the triangles, thus creating a curvature on the surface. The extension of the analysis in Chap. 7 to non-regular surfaces, such a blended triangular meshes, is left for future work.

References

1. Adams, R.J., Hannaford, B.: Stable haptic interaction with virtual environments. IEEE Trans. Robot. Autom. **15**(3), 465–474 (1999)
2. Cholewiak, S., Tan, H.Z.: Frequency analysis of the detectability of virtual haptic gratings. In: WHC '07: Proceedings of the Second Joint EuroHaptics Conference and Symposium on Haptic Interfaces for Virtual Environment and Teleoperator Systems, pp. 27–32. IEEE Computer Society, Washington (2007)
3. Hannaford, B., Ryu, J.H.: Time-domain passivity control of haptic interfaces. IEEE Trans. Robot. Autom. **18**(1), 1–10 (2002)
4. Kuchenbecker, K.J., Niemeyer, G.: Improving telerobotic touch via high-frequency acceleration matching. In: Proceedings of the IEEE Int. Conf. on Robotics and Automation (2006)
5. Pai, D.K., van den Doel, K., James, D.L., Lang, J., Lloyd, J.E., Richmond, J.L., Yau, S.H.: Scanning physical interaction behavior of 3D objects. In: SIGGRAPH '01: Proceedings of the 28th Annual Conference on Computer Graphics and Interactive Techniques, pp. 87–96. ACM Press, New York (2001)